SpringerBriefs in Materials

The SpringerBriefs Series in Materials presents highly relevant, concise monographs on a wide range of topics covering fundamental advances and new applications in the field. Areas of interest include topical information on innovative, structural and functional materials and composites as well as fundamental principles, physical properties, materials theory and design.

Indexed in Scopus (2022).

SpringerBriefs present succinct summaries of cutting-edge research and practical applications across a wide spectrum of fields. Featuring compact volumes of 50 to 125 pages, the series covers a range of content from professional to academic. Typical topics might include

- A timely report of state-of-the art analytical techniques
- A bridge between new research results, as published in journal articles, and a contextual literature review
- A snapshot of a hot or emerging topic
- An in-depth case study or clinical example
- A presentation of core concepts that students must understand in order to make independent contributions

Briefs are characterized by fast, global electronic dissemination, standard publishing contracts, standardized manuscript preparation and formatting guidelines, and expedited production schedules.

Masakazu Mukaida · Qingshuo Wei

Stable Organic Thermoelectric Devices and Thermocells for Self-powered Sensor Applications

Springer

Masakazu Mukaida
National Institute of Advanced Industrial Science and Technology (AIST)
Tsukuba, Ibaraki, Japan

Qingshuo Wei
National Institute of Advanced Industrial Science and Technology (AIST)
Tsukuba, Ibaraki, Japan

ISSN 2192-1091 ISSN 2192-1105 (electronic)
SpringerBriefs in Materials
ISBN 978-981-92-1037-4 ISBN 978-981-92-1038-1 (eBook)
https://doi.org/10.1007/978-981-92-1038-1

This Springer imprint is published by the registered company Springer Nature Singapore Pte Ltd.
The registered company address is: 152 Beach Road, #21-01/04 Gateway East, Singapore 189721, Singapore

Preface

There is a vast amount of unused waste heat all around us. From industrial pipes and server farms to the human body's warmth, this thermal energy is everywhere and represents a massive, untapped resource. If we could efficiently convert this low-grade heat into electricity, we could significantly improve global energy efficiency, reduce our reliance on fossil fuels and make a tangible impact on carbon dioxide emissions.

Meanwhile, the rapid global growth of the Internet of Things (IoT) creates an energy challenge. Although individual sensors used for monitoring factory environments, tracking logistics and providing continuous healthcare diagnostics consume very little power, their sheer number on a global scale will soon demand an enormous aggregate energy supply. Powering these trillions of sensors with conventional and disposable batteries is neither economically viable nor environmentally sustainable. Finding ways to autonomously meet this power demand is an urgent societal issue.

Historically, thermoelectric generators capable of harvesting this heat have relied on inorganic semiconductors. However, these materials are often rigid, expensive and sometimes toxic, which limits their seamless integration into our daily lives. Conversely, organic materials offer a highly promising, flexible and environmentally friendly alternative; however, they have traditionally been hindered by relatively low performance. There has long been a gap between the fundamental molecular science of organic thermoelectrics and their practical, system-level deployment.

We explore two primary, highly complementary pathways for harvesting low-temperature waste heat: solid-state organic thermoelectric conversion and liquid-state thermocells. Thanks to innovative device engineering and unique architectural designs, these carbon-based technologies have evolved to the point where they can serve as primary power sources for modern IoT sensors.

In the following chapters, we detail the technologies developed to reduce electrical resistance at material interfaces, and we explore the essential techniques employed to control and evaluate the structural anisotropy of organic thermoelectric materials. We then move on to liquid-state thermocells, detailing the development of quasi-solid-state gelled matrices and high-voltage modules.

This book provides comprehensive design guidelines, actual module structures and detailed analyses of circuits required to achieve impedance matching. We also explore the practical application of using these devices to charge lithium-ion batteries, demonstrating that these technologies can integrate seamlessly with existing electronic infrastructures.

This text is intended for materials scientists, electrical engineers and graduate students navigating the intersection of molecular design and energy harvesting. We hope that the research, technologies and system-level strategies introduced in these pages will inspire further cross-disciplinary innovation, reduce global energy waste and contribute meaningfully to a sustainable, low-carbon future.

Tsukuba, Japan

Masakazu Mukaida
Qingshuo Wei

Competing Interests The authors have no competing interests to declare that are relevant to the content of this manuscript.

Contents

Chapter 1
Introduction

Abstract The rapid global expansion of artificial intelligence (AI) and the Internet of Things (IoT) is driving a continuous surge in energy consumption and associated carbon dioxide emissions. Addressing this challenge requires moving beyond improvements in centralized power generation to develop sustainable, device-level energy supply solutions. Modern society produces vast amounts of untapped, low-temperature waste heat, with ubiquitous temperature gradients present in buildings, industrial equipment, outdoor infrastructure, and even the human body. This chapter introduces organic thermoelectric devices and thermocells as highly effective technologies for converting this ambient heat into electrical energy. By combining solid-state and liquid-state systems, thermal energy can be efficiently recovered across a wide temperature range—from room temperature up to approximately 200°C.

Keywords Organic thermoelectrics · Thermocells · Low-temperature waste heat recovery · Energy consumption

1.1 Why This Book

The rapid growth of global energy consumption, driven by electrification, digitalization, and the widespread deployment of artificial intelligence, continues to intensify concerns over carbon dioxide emissions and climate change [1]. While large-scale renewable energy technologies such as solar and wind power have expanded significantly, further reductions in carbon emissions will require not only improvements in centralized power generation but also innovations in how energy is supplied and consumed at the device level. In particular, a substantial fraction of energy in modern society is dissipated as low-grade waste heat, which remains largely unutilized.

In parallel, the proliferation of the Internet of Things (IoT) and distributed sensor networks is creating a strong demand for compact, lightweight, and maintenance-free power sources [2]. Sensors for environmental monitoring, infrastructure diagnostics, healthcare, and wearable electronics are often deployed in large numbers and in locations where battery replacement or recharging is impractical. Importantly, advances

M. Mukaida and Q. Wei, *Stable Organic Thermoelectric Devices and Thermocells for Self-powered Sensor Applications*, SpringerBriefs in Materials, https://doi.org/10.1007/978-981-92-1038-1_1

in low-power electronics and sensor technologies have significantly reduced the energy requirements of individual devices, enabling meaningful operation at power levels well below those traditionally associated with conventional power sources. This shift has opened new opportunities for energy-harvesting technologies that were previously considered impractical due to limited output power.

Thermal energy harvesting is particularly attractive in this context because temperature gradients are ubiquitous in both natural and engineered environments, including buildings, industrial equipment, outdoor infrastructure, and the human body [3–6]. Organic thermoelectric devices have attracted considerable attention as candidates for converting low-grade heat into electrical energy, owing to their mechanical flexibility, low thermal conductivity, solution processability, and compatibility with large-area fabrication [7–9]. Although the thermoelectric performance of organic materials remains inferior to that of state-of-the-art inorganic semiconductors, the electrical power generated by organic thermoelectric devices is increasingly well matched to the actual power demands of modern IoT sensors. As a result, the emphasis has gradually shifted from achieving record-high material figures of merit to realizing stable, reliable, and application-relevant device operation.

Complementary to solid-state organic thermoelectric devices, thermocells (thermogalvanic cells) represent another promising approach for harvesting thermal energy [10, 11]. Thermocells are liquid-based electrochemical systems rather than solid organic semiconductors, and their operating temperature range is therefore typically limited to relatively low temperatures, from room temperature to approximately 100 °C. Despite this constraint, thermocells can exhibit exceptionally large thermopower, often exceeding that of conventional thermoelectric materials by orders of magnitude. Such high Seebeck coefficients are particularly advantageous for driving power management circuits, including DC–DC converters, which require sufficient input voltage to operate efficiently under small temperature gradients [12].

Taken together, organic thermoelectric devices and thermocells offer complementary strengths for low-grade thermal energy harvesting. Organic thermoelectrics are well suited for solid-state integration and operation at moderate temperature gradients, while thermocells excel at low-temperature operation and high voltage output. By combining these two classes of devices, it becomes possible to harvest thermal energy across a broad temperature window, extending from room temperature to approximately 200 °C, which encompasses the majority of waste heat dispersed in everyday environments [13]. Such a hybrid perspective is particularly relevant for self-powered sensor systems, where stable operation, voltage compatibility, and long-term reliability are often more critical than maximizing peak power density.

In this book, we focus on stable organic thermoelectric devices and thermocells with an emphasis on their application in self-powered sensors. Rather than concentrating solely on material-level metrics, we critically examine stability challenges and design strategies spanning molecular engineering, morphology control, device architecture, and system integration.

It should be noted that while we have previously published review papers covering aspects of organic thermoelectrics and energy harvesting [14–17], this book serves as a more comprehensive and updated resource. Although some fundamental content may naturally overlap with our earlier work, we have made significant efforts to incorporate the latest research developments and our most recent experimental results. By highlighting both recent progress and remaining limitations, this work aims to clarify the pathways toward practical, reliable thermal energy harvesting for next-generation autonomous sensor platforms.

1.2 Thermoelectric Conversion for Waste-Heat Energy Harvesting

From the perspective of energy loss, waste heat represents one of the most significant and underutilized resources. In almost all human activities—including industrial manufacturing, transportation, and the operation of electrical appliances—input energy ultimately dissipates as heat and is released into the surrounding environment. Unlike electrical energy, which can be efficiently transmitted and stored, heat is difficult to confine and is inevitably lost through thermal diffusion. Although strategies such as thermal insulation, heat recovery, and heat storage have been widely explored, their effectiveness strongly depends on the temperature of the heat source [18]. High-temperature waste heat can be relatively easily reused, whereas low-temperature waste heat, characterized by small temperature differences, is often left unexploited due to poor efficiency [19].

In modern society, however, electricity remains the most versatile and demanded form of energy. Developing technologies that can directly convert low-grade waste heat into electrical power would therefore offer a powerful route toward improving overall energy efficiency and reducing carbon dioxide emissions. Such low-temperature heat sources are ubiquitous and include not only natural sources, such as solar irradiation, geothermal energy, and water temperature gradients, but also artificial sources such as factory exhaust, automobile components, electronic devices, and even the heat emitted by living organisms, including the human body. Harnessing these widely distributed thermal resources is particularly attractive for powering sensors and small electronic devices.

Power sources for sensors and Internet of Things (IoT) devices must satisfy stringent requirements, including compactness, low weight, and safety. For wearable and biomedical sensors, biocompatibility and mechanical compliance are essential, while for fixed sensors in factories, farms, or infrastructure, the available heat source is often dictated by the operating environment. In mobile or wearable systems, the heat source may be the device itself or the surrounding environment encountered during motion. These constraints have motivated increasing interest in solid-state energy-harvesting technologies that can operate quietly, reliably, and without mechanical components.

Thermoelectric conversion offers a direct, noise-free, and vibration-free method for converting heat into electricity and has been studied for nearly two centuries. The underlying mechanism originates from the Seebeck effect, discovered by Thomas

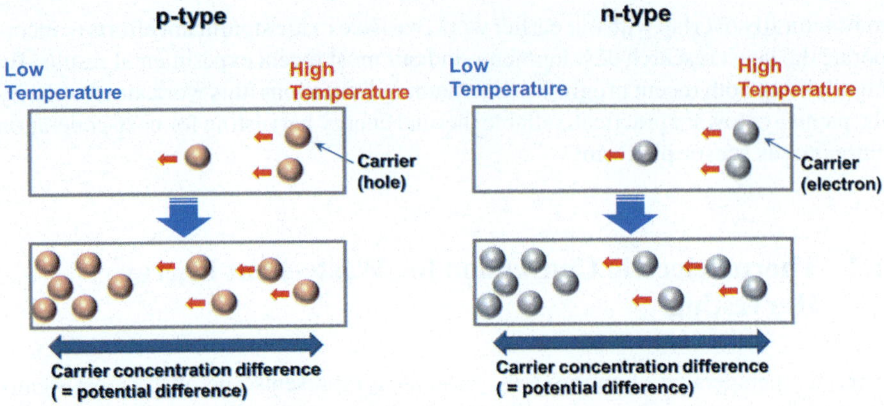

Fig. 1.1 Differences between thermoelectric phenomena of semiconductors in p-type and n-type

Johann Seebeck in 1821, in which a temperature difference across a material generates an electrical voltage due to the thermally driven motion of charge carriers [20]. In p-type semiconductors, holes migrate from the hot side to the cold side, resulting in a positive potential, whereas in n-type semiconductors, electrons serve as the dominant carriers, leading to a negative potential at the cold side as shown in Fig. 1.1. Although related thermoelectric phenomena such as the Peltier effects are also well known [21], this book focuses exclusively on thermoelectric power generation based on the Seebeck effect.

The performance of thermoelectric materials is commonly evaluated using three key parameters: the Seebeck coefficient (S), electrical conductivity (σ), and thermal conductivity (κ). A large Seebeck coefficient is desirable to generate high voltage from a given temperature difference, while high electrical conductivity minimizes resistive losses. At the same time, low thermal conductivity is essential to maintain a temperature gradient across the material. These factors are combined into the thermoelectric figure of merit, $Z = S^2\sigma/\kappa$, and its dimensionless form ZT, obtained by multiplying Z by the absolute temperature [22, 23]. The numerator, $S^2\sigma$, is often referred to as the power factor. While the thermoelectric figure of merit (ZT) and the power factor (σS^2) are useful metrics for comparing materials with similar thermal transport properties and operating conditions, they do not, by themselves, fully determine device-level performance. In practical thermoelectric generators, the achievable output power depends not only on σS^2 but also on the ability of the material to sustain a temperature gradient under heat flux. Materials with high thermal conductivity rapidly equilibrate the temperature between the hot and cold sides, severely limiting the effective ΔT across the device.

For example, many metals exhibit large power factors due to their high electrical conductivity; however, their intrinsically high thermal conductivity leads to substantial heat leakage, preventing the maintenance of a usable temperature difference. As a result, despite favorable σS^2 values, metals yield poor thermoelectric conversion efficiency and low power output at the device level.

Historically, thermoelectric research has focused primarily on inorganic semiconductors and ceramic materials [24], which are well suited for high-temperature operation and exhibit relatively high ZT values. However, the emergence of organic conductive materials with semiconducting electrical conductivity has opened new opportunities for thermoelectric energy harvesting [25–27]. Organic thermoelectric materials intrinsically possess low thermal conductivity and offer additional advantages such as mechanical flexibility, light weight, solution processability, and the use of abundant and potentially non-toxic elements. These characteristics make them particularly attractive for applications requiring conformability and safety, such as wearable electronics and distributed sensor networks. Although the thermoelectric performance of organic materials remains lower than that of state-of-the-art inorganic counterparts, their performance is increasingly well matched to the power requirements of modern low-power sensors, especially at temperatures below approximately 200 °C. Consequently, organic thermoelectrics are now being actively explored as practical energy-harvesting materials rather than competitors in high-temperature or high-power applications. A more detailed discussion of organic thermoelectric materials and their unique challenges will be presented in the following section.

In parallel with solid-state thermoelectric devices, thermocells—also referred to as thermogalvanic cells—offer an electrochemical route to thermal-to-electrical energy conversion. In thermocells, a temperature difference between two electrodes induces a voltage due to the temperature dependence of the redox potential of an electrochemical reaction as shown in Fig. 1.2. Fundamentally, this thermoelectric voltage originates from the entropy change associated with the redox reaction [28]. As a consequence, any redox couple that exhibits a temperature-dependent equilibrium potential can, in principle, function as a thermocell.

The thermopower (Seebeck coefficient) of a thermocell is determined by the reaction entropy and can be expressed as the temperature derivative of the redox potential. Redox couples with large entropy differences between their oxidized and reduced states can therefore generate exceptionally high thermopower, often exceeding that of solid-state thermoelectric materials by one or two orders of magnitude. This characteristic enables thermocells to produce relatively high output voltages even under small temperature gradients, which is particularly advantageous for use as power management circuits such as DC–DC converters in self-powered sensor systems.

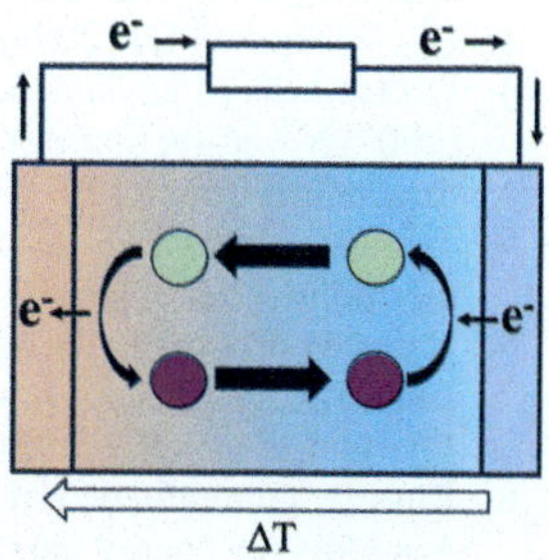

Fig. 1.2 A conceptual diagram of a typical thermocell. By applying a temperature difference to a structure in which a redox pair is sandwiched between electrodes, a continuous current is generated

For practical and sustainable operation, however, not all redox reactions are suitable for thermocells. To avoid material consumption at either the hot or cold electrode during continuous operation, both the oxidized and reduced species must be soluble in the electrolyte. This requirement ensures that redox reactions proceed reversibly and that mass transport between the two electrodes can maintain electrochemical equilibrium. If one of the redox states is insoluble or immobilized, irreversible deposition or depletion can occur, leading to electrode degradation and limited device lifetime. Consequently, the selection of redox couples for thermocells is strongly constrained by solubility, reversibility, and chemical stability considerations.

Due to their liquid-based nature, thermocells are generally limited to relatively low operating temperatures, typically from room temperature up to approximately 100 °C, depending on the volatility and stability of the electrolyte and redox species. Nevertheless, within this temperature window, thermocells offer distinct advantages, including large thermopower, stable voltage output, and flexibility. These features make them particularly well suited for harvesting low-grade thermal energy in environments where solid-state thermoelectric materials struggle to generate sufficient voltage.

Taken together, liquid-based thermocells and solid-state organic thermoelectric devices provide complementary strategies for low-grade thermal energy harvesting. Thermocells excel at low-temperature operation near room temperature and deliver high output voltage, while organic thermoelectric devices enable solid-state integration and operation at moderately elevated temperatures. By combining these two approaches, it becomes possible to harvest thermal energy across a broad temperature range—from room temperature to approximately 200 °C—which encompasses the majority of waste heat dispersed in everyday environments. Subsequent sections of this review will discuss the materials, device architectures, and stability challenges associated with both organic thermoelectric devices and thermocells in greater detail.

References

1. Lee H, Calvin K, Dasgupta D, Krinner G, Mukherji A, Thorne P, et al (2023) Climate change 2023: synthesis report, summary for policymakers. In: Core Writing Team, Lee H, Romero J (eds.) IPCC sixth assesment report. IPCC, Geneva, Switzerland, pp 1–34
2. Priya S, Inman DJ (2009) Energy harvesting technologies. Springer, New York
3. Dresselhaus MS, Chen G, Tang MY, Yang RG, Lee H, Wang DZ, Ren ZF, Fleurial JP, Gogna P (2007) New directions for low-dimensional thermoelectric materials. Adv Mater 19:1043–1053
4. Snyder GJ, Toberer ES (2008) Complex thermoelectric materials. Nat Mater 7:105–114
5. Tan GJ, Zhao LD, Kanatzidis MG (2016) Rationally designing high-performance bulk thermoelectric materials. Chem Rev 116:12123–12149
6. Haras M, Skotnicki T (2018) Thermoelectricity for IoT—a review. Nano Energy 54:461–476
7. Bubnova O, Crispin X (2012) Towards polymer-based organic thermoelectric generators. Energy Environ Sci 5:9345–9362
8. Russ B, Glaudell A, Urban JJ, Chabinyc ML, Segalman RA (2016) Organic thermoelectric materials for energy harvesting and temperature control. Nat Rev Mater 1:16050
9. Fan Z, Ouyang JY (2019) Thermoelectric properties of PEDOT:PSS. Adv Electron Mater 5:1800769

10. Zito R (1963) Thermogalvanic energy conversion. AIAA J 1:2133–2138
11. Quickenden TI, Mua Y (1995) A review of power-generation in aqueous thermogalvanic cells. J Electrochem Soc 142:3985–3994
12. Watson TC, Vincent JN, Lee H (2019) Effect of DC-DC voltage step-up converter impedance on thermoelectric energy harvester system design strategy. Appl Energ 239:898–907
13. Marina A, Spoelstra S, Zongtag HA, Wemmers AK (2021) An estimation of the European industrial heat pump market potential. Renew Sustain Energy Rev 139:110545
14. Wei Q, Mukaida M, Kirihara K, Naitoh Y, Ishida T (2015) Recent progress on PEDOT-based thermoelectric materials. Materials 8:732–750
15. Mukaida M, Kirihara K, Horike S, Wei Q (2020) Stable organic thermoelectric devices for self-powered sensor applications. J Mater Chem A8:22544–22556
16. Wei Q, Mukaida M, Kirihara K, Ishida T (2021) Thermoelectric materials-based on organic semiconductors. In: Funahashi R (ed) Thermoelectric energy conversion. Woodhead Publishing Oxford, pp 333–345
17. Wei Q, Horike S (2025) Advances in organic thermoelectric materials: from molecular design to device implementation. Inform Funct Mater 2:105–117
18. Ghadim HB, Godin A, Veillere A, Duquesne M, Haillot D (2025) Review of thermal management of electronics and phase change materials. Renew Sustain Energy Rev 208:115039
19. Tan ZM, Feng X, Wang YF (2021) Performance comparison of different heat pumps in low-temperature waste heat recovery. Renew Sustain Energy Rev 152:111634
20. Seebeck TJ (1826) Ueber die magnetische polarisation der metalle und erze durch temperatur-differenz. Annalen der Physik 82:133–160
21. DiSalvo FJ (1999) Thermoelectric cooling and power generation. Science 285:703–706
22. Ioffe AF (1957) Semiconductor thermoelements and thermoelectric cooling. Infosearch Limited, London
23. Rowe DM (2006) Thermoelectrics handbook. CRC Press, London
24. Mahan GD (1998) Good thermoelectrics. Solid State Phys 51:81–157
25. Fielding PE, Gutman F (1957) Electrical properties of phthalocyanines. J. Chem. Phys. 26:411–419
26. Pohl HA, Engelhardt EH (1962) Synthesis and characterization of some highly conjugated semiconducting polymers. J Phys Chem 66:2085–2095
27. Park YW, Denenstein A, Chiang CK, Heeger AJ, Macdiarmid AG (1979) Semiconductor-metal transition in doped (Ch)X—thermoelectric-power. Solid State Commun 29:747–751
28. Debethune AJ, Licht TS, Swendeman N (1959) The temperature coefficients of electrode potentials. J Electrochem Soc 106:616–625

Chapter 2
Organic Thermoelectric Materials

Abstract Analogous to inorganic semiconductors, organic thermoelectric materials are classified as either p-type or n-type. Historically, the development of n-type organic semiconductors lagged significantly behind p-type materials due to their inherent instability in ambient air. Among p-type materials, poly(3,4-ethylenedioxythiophene) (PEDOT) remains the most extensively studied, prized for its high electrical conductivity, low thermal conductivity, mechanical flexibility, and environmental compatibility. Beyond PEDOT, research is rapidly advancing on other p-type polymers which offer high crystallinity, long-range molecular ordering, and excellent charge carrier mobility. Crucially, the development of air-stable n-type organic thermoelectrics has recently been achieved with the introduction of poly-benzodifurandione (PBFDO). This advancement finally enables the fabrication of high-performance, all-organic p-n thermoelectric modules.

Keywords p-type · n-type · PEDOT · PBFDO · All organic

2.1 From Early Limitations to Practical Stability

Like inorganic thermoelectric semiconductors, organic thermoelectric materials are classified as p-type or n-type according to their majority charge carriers. However, for several decades following their initial exploration, organic thermoelectric materials were widely regarded as impractical because of their extremely low electrical conductivity and correspondingly poor thermoelectric performance.

In organic semiconductors, conventional Hall effect measurements are often difficult to apply due to low carrier mobility and the intrinsically disordered nature of these materials. As a result, the Seebeck coefficient has played a central role from the early stages of research as a practical parameter for identifying carrier types and estimating relative carrier concentrations. In 1957, Fielding et al. systematically investigated the Seebeck effect in metal-free phthalocyanines and copper phthalocyanine single crystals, concluding that holes are the majority carriers in both systems [1]. Later, Pohl et al. demonstrated that polyacene quinone radical polymers could

M. Mukaida and Q. Wei, *Stable Organic Thermoelectric Devices and Thermocells for Self-powered Sensor Applications*,
SpringerBriefs in Materials, https://doi.org/10.1007/978-981-92-1038-1_2

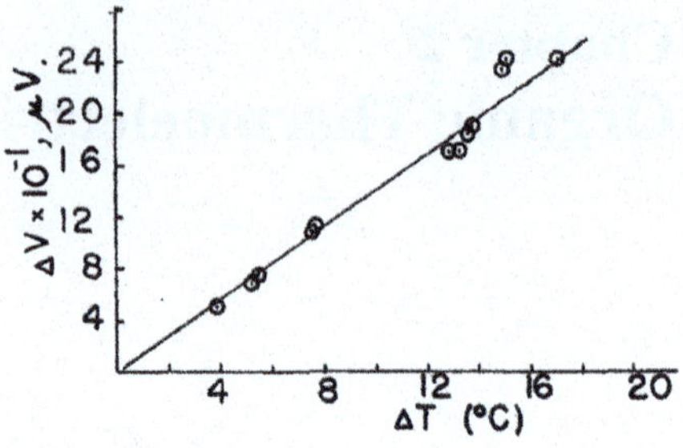

Fig. 2.1 Thermoelectric properties of polyacene quinone radical polymers [2]. Copyright 1962 American Chemical Society

exhibit either p-type or n-type behavior depending on their molecular structure [2], as inferred from the sign and magnitude of the Seebeck coefficient (Fig. 2.1).

The earliest quantitative evaluation of thermoelectric performance in organic materials dates back to 1964, when Katon et al. reported a thermoelectric figure of merit on the order of 10^{-8} K^{-1} for polyphthalocyanine, a nitrile-derived organic semiconductor polymer [3]. This value was several orders of magnitude lower than those typically required for practical inorganic thermoelectric materials, for which Z values exceeding ~10^{-3} K^{-1} are generally considered viable. Subsequent studies in the 1980s, such as the work by Park *et al.* on doped polyacetylene, reported moderate Seebeck coefficients (up to ~35 μV K^{-1} near room temperature) [4]. Nevertheless, the persistently low electrical conductivity of organic materials at the time reinforced the prevailing view that their thermoelectric performance would remain insufficient for practical applications.

Throughout the 1990s and early 2000s, research efforts focused primarily on understanding charge transport mechanisms rather than achieving competitive performance. Studies on perylene derivatives, polyaniline, and polypyrrole clarified hopping-dominated transport and low carrier mobility as fundamental constraints [5, 6]. A major conceptual shift occurred in the mid-2000s, when it became clear that organic thermoelectric materials should not be judged solely by the same benchmarks as inorganic semiconductors.

In 2007, Hiroshige *et al.* reported a ZT of approximately 0.1 at 313 K for phenylene vinylene derivatives, demonstrating that organic systems could achieve meaningful thermoelectric performance at low temperatures [7]. Importantly, this progress coincided with growing demand for lightweight, flexible, and low-toxicity power sources for emerging applications such as wearable electronics and distributed sensors.

Further momentum was generated in 2011, when Crispin *et al.* demonstrated that controlled dedoping of in situ polymerized, highly conductive poly(3,4-ethylenedioxythiophene):tosylate (PEDOT:tos) using tetrakis(dimethylamino)ethylene (TDAE) could yield a remarkably high power factor exceeding 300 μW m^{-1} K^{-2} [8]. Following these pioneering studies, the field of organic thermoelectrics expanded rapidly, encompassing materials design, device and module fabrication, and the development of dedicated characterization techniques [9, 10].

At the same time, it became evident that standardized characterization protocols for organic thermoelectric materials were still evolving. Factors such as anisotropic thermal transport in polymer films and variability in Seebeck coefficient measurements between laboratories have complicated direct performance comparisons. In particular, for thin-film samples, the details of thermocouple–film contact and temperature gradient definition can significantly influence measured values. Addressing these measurement challenges has since become an important focus of the field and has contributed to the maturation and increasing reliability of organic thermoelectric research [11–14].

2.2 PEDOT-Based p-Type Thermoelectric Materials

Among organic thermoelectric materials, poly(3,4-ethylenedioxythiophene) (PEDOT) (Fig. 2.2) has emerged as the most extensively studied p-type system [15–17]. This prominence is due to its unique combination of high electrical conductivity, intrinsically low thermal conductivity, mechanical flexibility, and environmental compatibility.

In practical applications, PEDOT is most commonly processed as a composite with polystyrene sulfonate (PSS), forming PEDOT:PSS. This formulation offers excellent aqueous dispersibility and compatibility with scalable, solution-based thin-film fabrication techniques such as inkjet printing and roll-to-roll processing.

The thermoelectric properties of PEDOT: PSS are governed primarily by its nanoscale morphology. While PSS is essential for stabilizing PEDOT colloids in water, it is an electrically insulating component. If PSS forms poorly connected or excessive domains, it can significantly hinder charge transport. During film formation, the rapid evaporation of solvent from aqueous dispersions often results in "inetically trapped" morphologies. In these states, PEDOT-rich regions are partially isolated by disordered PSS, which limits the creation of effective charge carrier pathways. A major breakthrough in PEDOT: PSS thermoelectrics was the discovery that incorporating high-boiling-point polar additives—such as ethylene glycol (EG) or dimethyl sulfoxide (DMSO)—into the casting solution can dramatically enhance electrical conductivity [18–21]. These additives modify the drying kinetics and screen the Coulombic interactions between the PEDOT and PSS chains, promoting phase separation and structural reorganization.

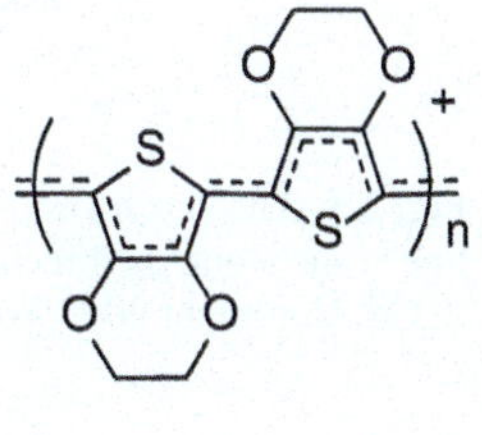

Fig. 2.2 Chemical structure of doped poly(3,4-ethylenedioxythiophene)

As a result of this treatment, PEDOT chains adopt a more extended conformation and pack into better-connected, percolated networks. Grazing-incidence small-angle X-ray scattering (GISAXS) and related structural probes have revealed enhanced lamellar ordering and increased crystallinity in these additive-treated films. This structural refinement correlates directly with improved carrier mobility, as illustrated in Figs. 2.3 and 2.4.

Importantly, this morphology optimization does more than just increase electrical conductivity; it also introduces pronounced structural anisotropy within the films. The preferential in-plane alignment of PEDOT-rich domains leads to direction-dependent charge and heat transport.

This anisotropy has direct implications for how we interpret thermoelectric parameters, including electrical conductivity (σ), thermal conductivity (κ), and the derived figure of merit (ZT). Consequently, reported performance metrics for PEDOT: PSS—and organic thermoelectric materials more broadly—are inherently linked to the specific measurement direction and device geometry employed during characterization.

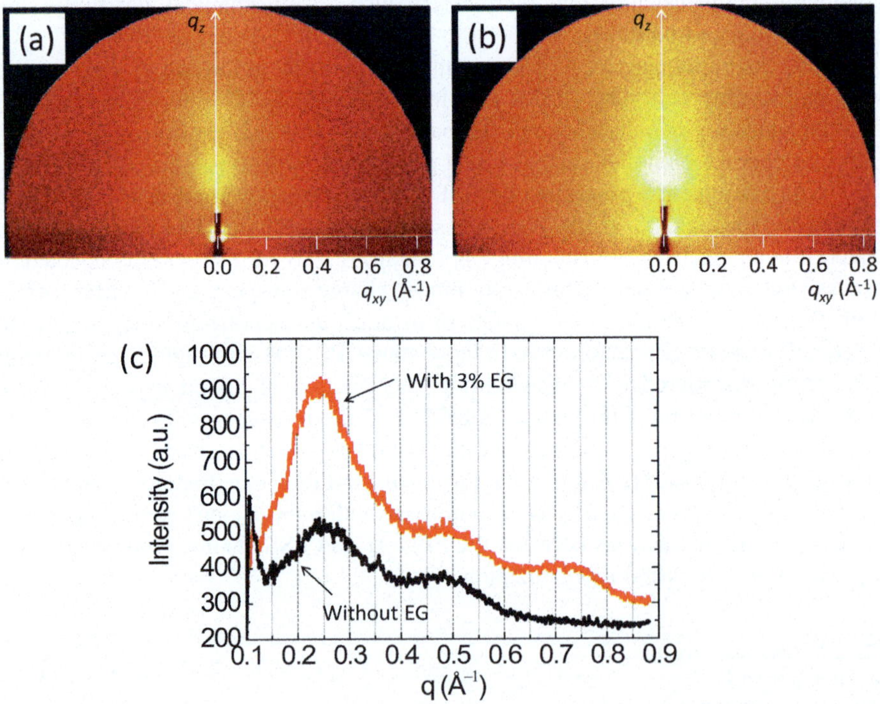

Fig. 2.3 The 2D GISAXS patterns of the PEDOT: PSS film: **a** without EG, **b** with 3% EG. **c** Detailed line scans along q_z direction. Ther incident angle was fixed at 0.5°and the exposure time was 900 s [18]. Copyright 2013 Wiley

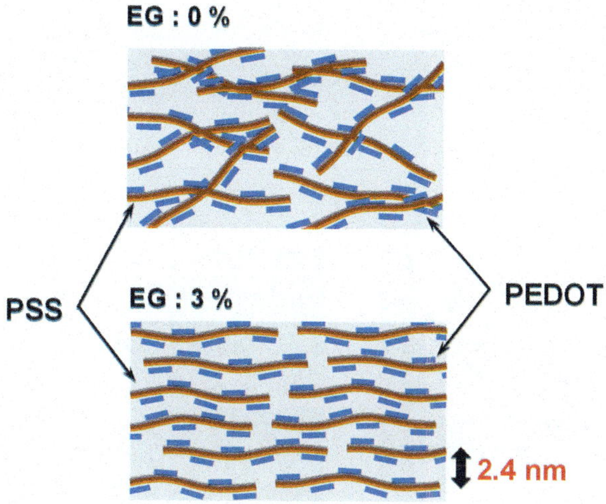

Fig. 2.4 Effect of EG addition on structures of PEDOT: PSS

Through these morphology control strategies, the production of PEDOT: PSS films with electrical conductivities exceeding 10^3 S cm^{-1} has become routinely achievable, all while preserving the intrinsically low thermal conductivity characteristic of organic polymers.

A landmark study by Park et al. (Fig. 2.5) demonstrated that solution-processed PEDOT: PSS films can achieve exceptionally high thermoelectric performance through sequential strong acid–base treatments [22]. By utilizing trifluoromethanesulfonic acid (TFSA) to remove excess PSS and induce highly crystalline PEDOT domains, followed by precise de-doping using vapor-phase tetrakis(dimethylamino)ethylene (TDAE), they realized a record PF of 534.5 μW m^{-1} K^{-2}.

Notably, this treatment strategy encouraged molecular alignment and enhanced π–π stacking within the film (Fig. 2.5b), thereby maintaining delocalized charge transport characteristics. These findings underscore the critical necessity of maintaining a delicate balance between carrier concentration and structural connectivity to achieve high-performance PEDOT-based thermoelectric materials.

2.3 Other p-Type Thermoelectric Materials

Many efforts have been directed toward advancing p-type semiconducting polymers beyond PEDOT-based systems, with a particular focus on materials such as poly(2,5-bis(3-alkylthiophen-2-yl)thieno[3,2-b]thiophene) (PBTTT) and its derivatives [23–28]. These materials are valued for their high crystallinity, long-range molecular ordering, and exceptional charge carrier mobility.

In their pristine state, PBTTT polymers are typically undoped, but they can be effectively p-doped using strong molecular oxidants like F_4TCNQ or $FeCl_3$. The

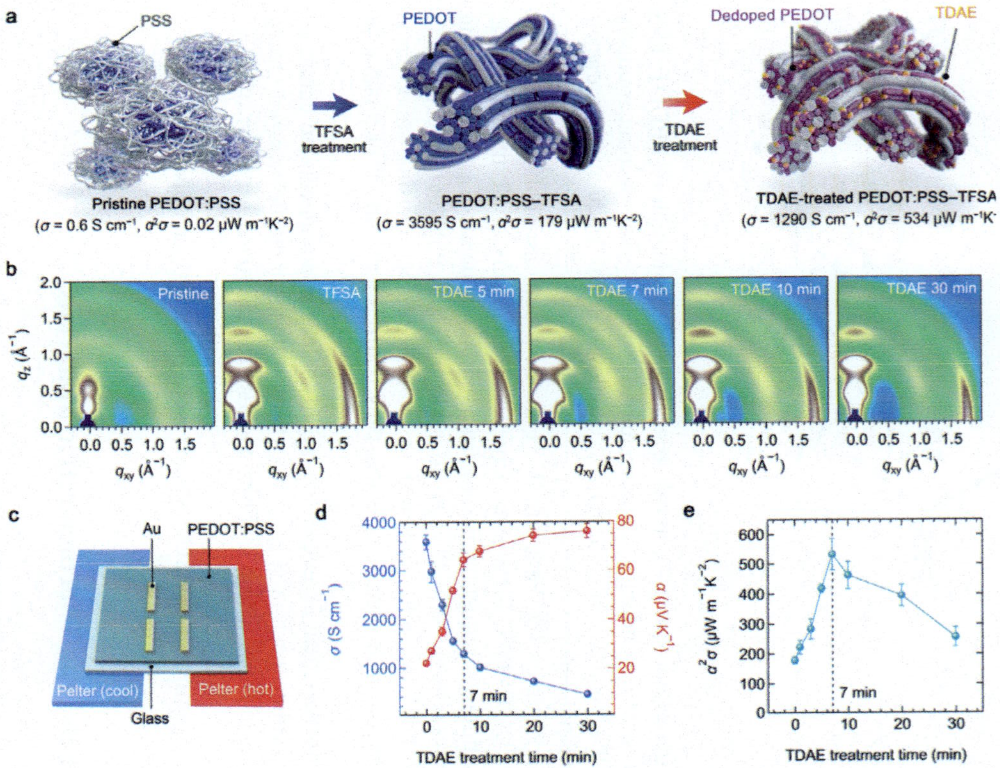

Fig. 2.5 Thermoelectric and structural properties of TDAE-treated PEDOT: PSS–TFSA films. **a** Schematic illustrations of PEDOT and PSS networks in the pristine, PEDOT: PSS–TFSA, and TDAE-treated PEDOT: PSS–TFSA films. **b** 2D GIWAXS patterns of the films along with the TDAE-treatment time. **c** Illustrative structure of the TE devices and measurement. Thermoelectric performance of the PEDOT: PSS–TFSA films in terms of **d** electrical conductivity (σ), Seebeck coefficient (α), and **e** power factor (PF) as a function of the TDAE-treatment time [22]. Copyright 2024 Wiley

crystalline nature of PBTTT allows these dopant molecules to intercalate into the polymer layers, significantly enhancing electrical conductivity while maintaining the structural integrity required for efficient charge transport.

One of the key advances in this area was the implementation of vapor-phase doping. Patel et al. [23, 24] reported that vapor-doped PBTTT films achieved charge carrier mobilities exceeding 2 $cm^2\ V^{-1}\ s^{-1}$, electrical conductivities up to 670 S cm^{-1}, and power factors as high as 120 $\mu W\ m^{-1}\ K^{-2}$. These improvements were attributed to more effective dopant infiltration and reduced disruption to the polymer's semicrystalline morphology compared to solution-based methods.

In subsequent research, the same group demonstrated that the orientational correlation length (OCL)—rather than the absolute crystallite size—correlates more strongly with charge transport properties, establishing it as a crucial parameter for thermoelectric optimization. Building on this, the use of a weaker electron acceptor, F2TCNQ, also demonstrated surprising efficacy in the vapor-phase doping of PBTTT. Despite a less favorable energetic alignment for charge transfer, F2TCNQ-doped PBTTT achieved a power factor of approximately 70 $\mu W\ m^{-1}\ K^{-2}$. This performance was attributed to improved dopant miscibility and minimal disruption to the polymer microstructure; the weaker interaction allowed the polymer to maintain superior chain packing and orientation, supporting high charge mobility even at moderate doping levels.

Chen et al. further investigated ion-exchange doping in PBTTT using a series of closed-shell anions (TFO^-, $TFSI^-$, $TFSM^-$) to explore how the counterion size affects thermoelectric transport [29]. They found that at intermediate conductivities (~0.1–1 S cm^{-1}), the Seebeck coefficient and conductivity remained broadly independent of the ion size. This was rationalized by a heterogeneous doping mechanism in which amorphous domains are doped preferentially. In this regime, polaron delocalization becomes sufficient to average out Coulombic interactions, thereby reducing the influence of counterion size on transport.

At higher conductivities (>100 S cm^{-1}), doping extends into crystalline domains, which enables a transition to a metal-like transport regime. Maximum power factors exceeding 30 $\mu W\ m^{-1}\ K^{-2}$ were achieved, which reinforced the importance of controlled doping strategies and morphology optimization in emerging p-type organic thermoelectric materials.

Conjugated donor–acceptor (D–A) polymers have emerged as another promising class of thermoelectric materials due to their tunable energy levels, narrow bandgaps, and favorable charge transport properties. Unlike homopolymers, the alternating electron-rich (donor) and electron-deficient (acceptor) units in these systems allow for precise control over the electronic structure [30].

For example, Ding et al. demonstrated that backbone engineering of naphthobisthiadiazole (NTz)-based D–A polymers through the incorporation of various thiophene-based donor units can effectively modulate molecular packing, dopant miscibility, and charge carrier mobility [31]. By incorporating various thiophene-based donor units, they were able to tailor the polymer's interaction with dopants.

In particular, a quaterthiophene-based polymer (PNTz4T) featuring a linear backbone achieved a peak power factor (PF) exceeding 30 $\mu W\ m^{-1}\ K^{-2}$ (Fig. 2.6). This

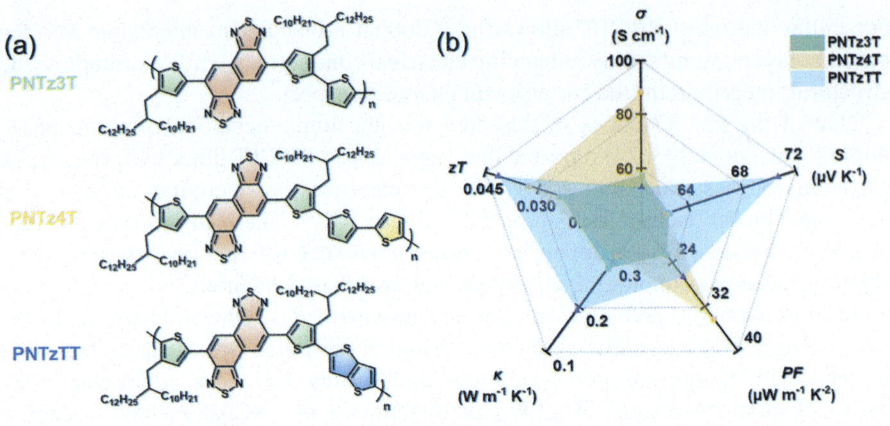

Fig. 2.6 **a** Chemical structures of NTz-based D-A conjugated polymers; **b** Chart of σ, S, PF, κ, and zT of doped PNTz3T, PNTz4T, and PNTzTT thin films. Reproduced with permission from ref [31]. Copyright 2024 Wiley

high performance was driven by enhanced edge-on stacking and efficient dopant intercalation within the polymer matrix. These findings underscore the potential of rational molecular design in optimizing both electrical conductivity and the Seebeck coefficient in D–A systems, even in materials that lack long-range crystallinity.

A major conceptual breakthrough was achieved by Wang et al., who introduced a polymeric multi-heterojunction (PMHJ) strategy [32]. This approach was designed specifically to overcome the longstanding trade-off between electrical conductivity and thermal conductivity in organic thermoelectric materials.

The PMHJ structure consists of alternating nanoscale layers of two high-mobility polymers: the DPP-based donor–acceptor polymer PDPPSe-12 and PBTTT. These layers are separated by gently interpenetrating bulk heterojunction interfaces, forming a quasi-periodic architecture with sub-10-nm spacing.

This sophisticated design enhances interfacial phonon scattering, which significantly lowers the in-plane thermal conductivity ($\kappa\|$), while maintaining efficient charge transport through the high-mobility pathways. Through this method, they achieved a record ZT of greater than 1 at 368 K, surpassing the performance of both commercial inorganic thermoelectric materials and previously reported organic systems in that temperature range.

Their work challenges the long-held assumption that high crystallinity is the only route to high performance. Instead, it demonstrates that functionally engineered disorder and hierarchical heterojunctions can successfully decouple electrical conductivity, the Seebeck coefficient, and thermal conductivity. This creates a new class of molecular "thermoelectric plastics" optimized for flexible, scalable, and high-efficiency energy harvesting.

These collective advances reflect a fundamental shift in the research of p-type organic thermoelectric materials. The focus has evolved from the initial morphology engineering of PEDOT derivatives to sophisticated molecular doping and band engineering, and finally toward emerging concepts of functional disorder and polar energy landscapes. These strategies enable the development of scalable, high-performance materials capable of operating efficiently under low-temperature gradients—an essential requirement for flexible and wearable applications.

2.4 n-Type Organic Thermoelectric Materials

While significant advances had been achieved in p-type organic thermoelectric materials, the lack of stable n-type counterparts long represented a fundamental bottleneck for the field [33]. Efficient thermoelectric modules require both p-type and n-type legs to maximize voltage output and minimize parasitic losses.

However, most early n-type organic semiconductors exhibited severe air instability. Upon exposure to oxygen or moisture, their electrical conductivity rapidly decayed—often within minutes [34]. This degradation was due to a combination of factors: inefficient electron doping, shallow lowest unoccupied molecular orbital (LUMO) energy levels, and the intrinsic instability of radical anions in ambient conditions. Consequently, reported n-type conductivities were typically limited to the range of 0.1–100 S cm^{-1} and could only be maintained under inert atmospheres. This stood in sharp contrast to the long-term stability and high conductivity ($>10^3$ S cm^{-1}) routinely achieved in p-type systems.

Subsequent studies have further clarified the molecular and material design principles required for stable n-type systems. These include:

(i) Electronic Stabilization: Stabilizing radical anions through deep LUMO levels (< -4.0 eV) and rigid backbones.
(ii) Structural Optimization: Minimizing structural disorder to enhance charge carrier mobility.
(iii) Doping Control: Controlling doping sites and morphology to suppress dedoping and phase segregation.

Early efforts in this space relied heavily on fullerene derivatives, such as [6,6]-Phenyl-C61-Butyric Acid Methyl Ester (PCBM). These materials are attractive because they possess an inherently low thermal conductivity (~0.1 W m^{-1} K^{-1}), a result of their "molecular phonon–glass" nature—meaning they scatter heat as effectively as glass while maintaining the potential for electronic transport [34]. However, the performance of early PCBM systems was severely limited by low doping efficiency and poor miscibility with n-type dopants.

A major advancement was the development of 1,3-dimethyl-2-phenyl-2,3-dihydro-1H-benzoimidazole (DMBI) derivatives. These function as solution-processable dopants that generate neutral radicals, enabling efficient electron transfer. Wei et al. [35] demonstrated that n-DMBI could significantly enhance both the conductivity and air stability of PCBM-based n-channel organic transistors, marking one of the first successful applications of molecular n-doping in solution-processed systems.

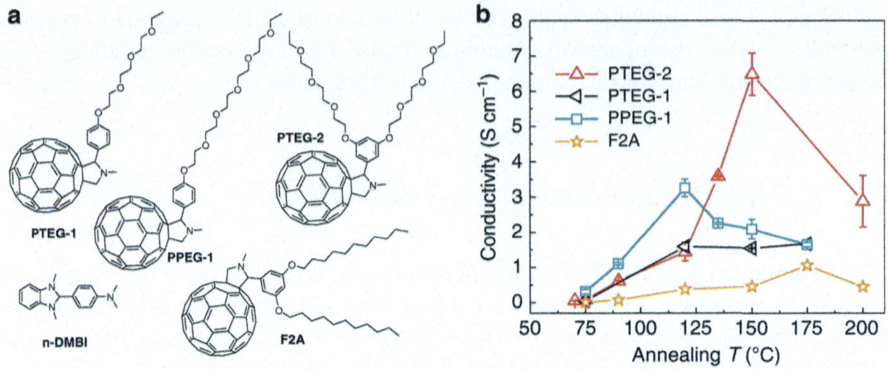

Fig. 2.7 **a** Chemical structures of different fullerene derivatives (PTEG-1, PTEG-2, PPEG-1, and F2A), and dopant (n-DMBI); **b** Plots of electrical conductivity at room temperature as a function of the annealing temperature for different fullerene derivatives doped at a concentration of 8 wt% n-DMBI. Error bars indicate the standard errors of the mean values of electrical conductivity obtained by the measurement of six different samples. Reproduced with permission from ref. [36]. Copyright 2020 Springer Nature

Liu et al. achieved a significant breakthrough in fullerene-based n-type materials through precise side-chain engineering [36]. As illustrated in Fig. 2.7, they designed a fullerene derivative (PTEG-2) incorporating arm-shaped, double-triethylene-glycol side chains. These chains were specifically engineered to improve dopant miscibility and induce a "disorder-to-order" transition upon thermal annealing.

This rational molecular design facilitated two critical outcomes: a high doping efficiency of approximately 60% and a well-ordered packing structure. These factors combined to produce an electrical conductivity exceeding 10 S cm^{-1} and an ultralow thermal conductivity of less than 0.1 W m^{-1} K^{-1}. The resulting ZT reached a record 0.34 at 120 °C. This achievement represents a rare and successful example of an organic thermoelectric material approaching the Phonon–Glass Electron Crystal (PGEC) ideal, where thermal transport is suppressed without compromising electronic mobility.

In parallel with the development of small-molecule systems, significant progress has been made in n-type polymers, particularly those based on naphthalenediimide (NDI) [37–39]. Early representatives such as P(NDI2OD-T2), commonly known as N2200, showed initial promise due to their deep lowest unoccupied molecular orbital (LUMO) levels and high thermal stability.

However, these polymers initially suffered from low doping efficiency and significant phase separation when mixed with common n-type dopants like n-DMBI. These morphological issues resulted in limited electrical conductivity and poor air stability. Wang et al. addressed these constraints by introducing a more planar bithiazole comonomer into the polymer backbone [40]. This modification yielded a copolymer with improved molecular packing and enhanced dopant accessibility, resulting in an electrical conductivity of ~0.1 S cm^{-1} and PF of 1.5 μW m^{-1} K^{-2} —values more than two orders of magnitude higher than those of the parent polymer.

Further advancements in this class of materials focused on optimizing dopant miscibility through side-chain engineering. Liu et al. replaced the hydrophobic alkyl chains of N2200 with polar triethylene glycol (TEG)-based side chains to produce a derivative termed TEG-N2200 [41].

The introduction of these polar side chains dramatically improved the miscibility between the polymer matrix and the dopant. This structural optimization enabled the doped TEG-N2200 to achieve a significantly enhanced electrical conductivity of 0.17 S cm^{-1}, reinforcing the critical role of side-chain polarity in the development of efficient n-type organic thermoelectrics.

The development of thermally activated and highly miscible dopants, such as triaminomethane (TAM), has been pivotal to the advancement of n-type organic thermoelectric materials [42]. TAM exhibits exceptional chemical stability—remaining stable even in boiling water and concentrated hydrochloric acid—and demonstrates a strong hydride-donating ability upon thermal activation. As shown in Fig. 2.8, its molecular structure is characterized by a guanidine-type cation with a small molecular volume. This compact structure enhances miscibility with polymer side chains, facilitating uniform doping and minimizing disruption to the critical π–π stacking of the polymer backbone.

When utilized to dope fluorinated benzodifurandione-based poly(phenylene vinylene) (FBDPPV), a 2020 study demonstrated that TAM can help achieve electrical conductivities of up to 21 S cm^{-1} and power factors (PF) of up to 51 $\mu W\ m^{-1}\ K^{-2}$. Remarkably, these metrics were maintained even in films exceeding 10 μm in thickness, ranking among the highest values reported for n-type organic polymers at that time.

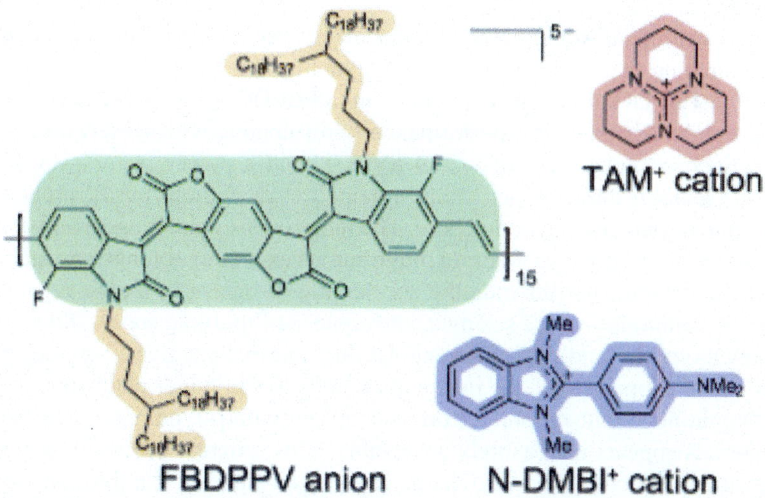

Fig. 2.8 Chemical structures of FBDPPV anions, TAM^+, and N-$DMBI^+$ cations [42]. Copyright 2020 Springer Nature

The thermally activated doping process of TAM is particularly advantageous because it mitigates phase separation and structural disruption, ensuring a uniform morphology and high stability within the doped films. Furthermore, the high solubility of TAM in common organic solvents facilitates the fabrication of thick, uniform films—a key requirement for the practical application of thermoelectric devices. The combination of stability, robust doping capability, and compatibility with polymer matrices makes TAM a major milestone in the pursuit of high-performance n-type organic thermoelectrics.

Among n-type polymers, PBFDO (poly(benzodifurandione)) stands as one of the most promising thermoelectric materials currently available. Synthesized via oxidative polymerization and *in situ* n-doping, PBFDO exhibits an exceptional electrical conductivity exceeding 2000 S cm^{-1}—ranking among the highest reported values for n-type conjugated polymers as of 2022. With a Seebeck coefficient of ~21 μV K^{-1} and a power factor (PF) exceeding 50 μW m^{-1} K^{-2}, it is uniquely suited for high-performance organic electronics [43].

Recent research has expanded on this platform. Ke et al. [44] introduced a related n-PBDF system synthesized via copper-catalyzed oxidative polymerization under ambient conditions. This approach resulted in high conductivity combined with excellent thermal and environmental stability across a broad temperature range from −20 to 65 °C.

Most recently, Huang et al. developed a novel one-step synthesis method for PBFDO using α-tocopherylquinone (α-TQ) as a catalyst (Fig. 2.9) [45]. This represents a significant improvement over the traditional tetramethylquinone (TMQ) route. While the TMQ route often suffers from catalyst crystallization and aggregation, α-TQ features a bulky, saturated side chain that suppresses crystallization upon reduction.

This structural advantage ensures that α-TQ remains well-dispersed in solution, eliminating the need for dialysis and enabling the direct production of highly conductive PBFDO films from a stable ink. PBFDO synthesized via the α-TQ method exhibits outstanding performance, with conductivities exceeding 1300 S cm^{-1}, a Seebeck coefficient of −30.9 μV K^{-1}, and a power factor over 100 μW m^{-1} K^{-2}. Notably, these properties are retained even without purification, significantly reducing processing complexity. The material also maintains over 97% of its conductivity after 180 days in air, highlighting its exceptional long-term stability.

These developments illustrate the accelerating progress in n-type organic thermoelectric materials. We are seeing a convergence of multiple successful strategies, including molecular and side-chain engineering, thermally activated doping, and the emergence of "dopant-free" yet inherently conductive polymer platforms.

With continued innovation, performance parity with p-type organic thermoelectric materials appears increasingly achievable. This progress paves the way for all-organic thermoelectric modules characterized by balanced, durable, and scalable performance, moving the technology closer to practical, real-world deployment.

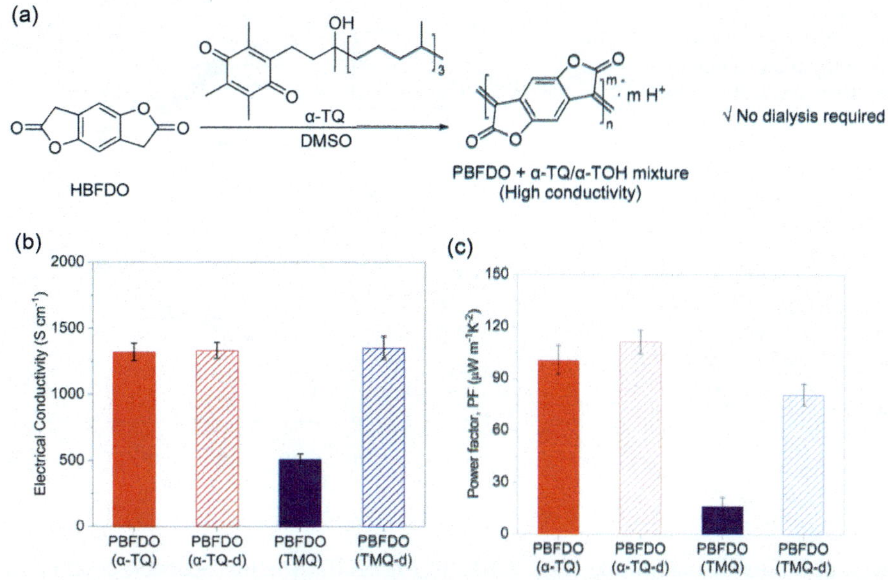

Fig. 2.9 **a** Schematic diagram of PBFDO synthesis using α-TQ; **b** Electrical conductivity and **c** power factor (PF) of α-TQ-synthesized PBFDO (before and after dialysis) and TMQ-synthesized PBFDO (before and after dialysis) [45] Copyright 2025 Wiley

2.5 Performance Optimization

Optimizing the thermoelectric performance of organic materials typically centers around enhancing the power factor (= $S^2\sigma$) while maintaining the low thermal conductivity inherent to polymers. A fundamental challenge in this pursuit is the precise control of the doping level. Unlike inorganic semiconductors, where carrier concentration can be tuned accurately through chemical or electrostatic doping, organic semiconductors typically rely on redox reactions involving molecular dopants.

Consequently, controlling the exact number and spatial distribution of charge carriers is significantly more difficult. In many cases, the polymer exists in either a highly doped or nearly undoped state, and the situation is further complicated by dopant diffusion and induced morphological changes. Despite significant research progress, achieving the same level of precise, reproducible control over carrier density that is standard in inorganic systems remains limited in the organic domain.

Despite these challenges, numerous studies—including the seminal work by Glaudell et al. [46]—have shown that increasing dopant density generally enhances thermoelectric performance. Specifically, while the electrical conductivity (σ) increases exponentially with the molar ratio of dopant to polymer, the Seebeck coefficient (S) decreases sub-linearly. This results in a PF that increases with doping over

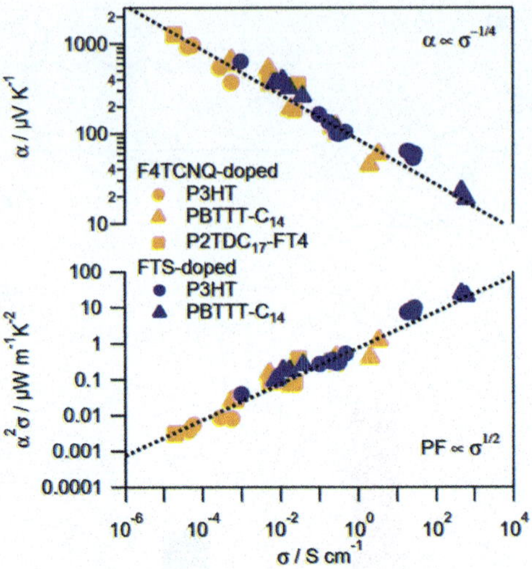

Fig. 2.10 Thermopower (top) and power factor (bottom) as a function of electrical conductivity for F4TCNQ-doped polymers (yellow) P3HT (●), PBTTT-C_{14} (▲), and P2TDC$_{17}$-FT4 (·) and FTS-doped polymers (blue) P3HT (●), and PBTTT-C_{14} (▲). The dashed lines indicate an empirical fit of α proportional to $\sigma^{-1/4}$ and PF proportional to $\sigma^{1/2}$ [46]. Copyright 2014 Wiley

several orders of magnitude (Fig. 2.10). In Fig. 2.10, α is the Seebeck coefficient. Note that while the Seebeck coefficient is often denoted by S, it can also be written as α.

This monotonic trend contrasts with inorganic systems, where the carrier concentration can typically be optimized at a specific density to maximize both PF and *ZT*. As shown in Fig. 2.10, Glaudell et al. established that the empirical relationships $S \propto \sigma^{-1/4}$ and $PF \propto \sigma^{1/2}$ held consistently across various doping strategies and semiconducting polymers. Their data suggested that a maximum PF was not reached even at the highest measured conductivities of ~600 S cm^{-1}.

In 2016, we introduced a photoinduced de-doping strategy for PEDOT: PSS that enabled continuous and precise control over the carrier density across more than three orders of magnitude [47]. As shown in Fig. 2.11, this method allowed us to probe the full range of doping-dependent transport properties, which revealed an optimal carrier density (~5×10^{20} cm^{-3}) at which the Seebeck coefficient and electrical conductivity together produced the highest PF (~42 μW m^{-1} K^{-2}).

At higher doping levels, corresponding to conductivities exceeding 600 S/cm, the Seebeck coefficient was observed to drop more rapidly than the electrical conductivity increased, leading to a subsequent decline in the PF. These findings highlight the critical necessity of precise and tunable doping strategies for organic thermoelectric materials.

The results suggest that with sufficient control over carrier concentration—though such precise, reproducible control remains limited across the broader field compared to inorganic standards—organic thermoelectric materials can exhibit performance optimization profiles similar to those of their inorganic counterparts. This realization paves the way for a more rational design approach to achieve high-efficiency energy harvesting.

Fig. 2.11 **a** Electromotive force as a function of the temperature difference for obtaining the Seebeck coefficient of a PEDOT:PSS film at a given ultraviolet irradiation time, **b** resulting Seebeck coefficient, and **c** power factor as a function of the carrier density [47]. Copyright 2016 American Chemical Society

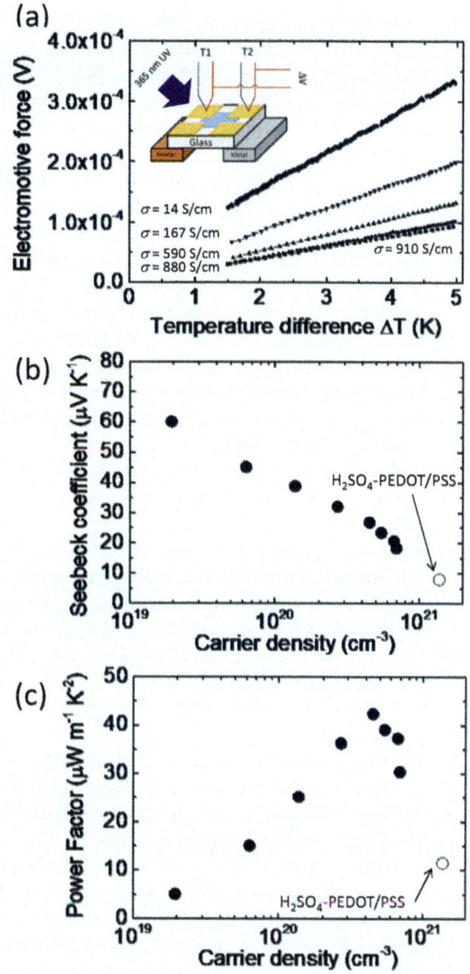

References

1. Fielding PE, Gutman F (1957) Electrical properties of phthalocyanines. J Chem Phys 26:411–419
2. Pohl HA, Engelhardt EH (1962) Synthesis and characterization of some highly conjugated semiconducting polymers. J Phys Chem 66:2085–2095
3. Katon JE, Wildi BS (1964) Semiconducting organic polymers derived from nitriles. Thermoelectric power and thermal conductivity measurements. J Chem Phys 40:2977–2982
4. Park YW, Han YK, Choi CH, Shirakawa H (1984) Metallic nature of heavily doped polyacetylene derivatives: thermopower. Phys Rev B 30:5847–5851
5. Meyer JP, Schlettwein D, Wöhrle D, Jaeger NI (1995) Charge transport in thin films of molecular semiconductors as investigated by measurements of thermoelectric power and electrical conductivity. Thin Solid Films 258:317–324

6. Mateeva N, Niculescu H, Schlenoff J, Testardi LR (1998) Correlation of Seebeck coefficient and electric conductivity in polyaniline and polypyrrole. J Appl Phys 83:3111–3117
7. Hiroshige Y, Ookawa M, Toshima N (2007) Thermoelectric figure-of-merit of iodine-doped copolymer of phenylenevinylene with dialkoxyphenylenevinylene. Synth Met 157:467–474
8. Bubnova O, Khan ZU, Malti A, Braun S, Fahlman M, Berggren M, Crispin X (2011) Optimization of the thermoelectric figure of merit in the conducting polymer poly(3,4-ethylenedioxythiophene). Nat Mater 10:429–433
9. Bubnova O, Khan ZU, Wang H, Braun S, Evans DR, Fabretto M et al (2014) Semi-metallic polymers. Nat Mater 13(2014):190–194
10. Massonnet N, Carella A, Jaudouin O, Rannou P, Laval G, Celle C, Simonato JP (2014) Improvement of the Seebeck coefficient of PEDOT:PSS by chemical reduction combined with a novel method for its transfer using free-standing thin films. J Mater Chem C2:1278–1283
11. Wei Q, Mukaida M, Kirihara K, Ishida T (2014) Experimental studies on the anisotropic thermoelectric properties of conducting polymer films. Acs Macro Lett 3:948–952
12. Weathers A, Khan ZU, Brooke R, Evans D, Pettes MT, Andreasen JW et al (2015) Significant electronic thermal transport in the conducting polymer Poly(3,4-ethylenedioxythiophene). Adv Mater 27:2101–2106
13. Liu J, Wang XJ, Li DY, Coates NE, Segalman RA, Cahill DG (2015) Thermal conductivity and elastic constants of PEDOT: PSS with high electrical conductivity. Macromolecules 48:585–591
14. Khan ZU, Edberg J, Hamedi MM, Gabrielsson R, Granberg H, Wagberg L, et al (2016) Thermoelectric polymers and their elastic aerogels. Adv Mater 28:4556–562
15. Elshner A, Kirchmeyer A, Lovenich W, Merker U, Reuter K (2010) PEDOT: principles and applications of an intrinsically conductive polymer. CRC Press, Boca Raton
16. Lövenich W (2014) PEDOT-properties and applications. Polym Sci Ser C 56:135–143
17. Sun K, Zhang SP, Li PC, Xia YJ, Zhang X, Du DH et al (2015) Review on application of PEDOTs and PEDOT: PSS in energy conversion and storage devices. J Mater Sci-Mater El 26:4438–4462
18. Wei Q, Mukaida M, Naitoh Y, Ishida T (2013) Morphological change and mobility enhancement in PEDOT: PSS by adding co-solvents. Adv Mater 25:2831–2836
19. Takano T, Masunaga H, Fujiwara A, Okuzaki H, Sasaki T (2012) PEDOT nanocrystal in highly conductive PEDOT: PSS polymer films. Macromolecules 45:3859–3865
20. Ouyang J, Xu QF, Chu CW, Yang Y, Li G, Shinar J (2004) On the mechanism of conductivity enhancement in poly (3,4-ethylenedioxythiophene): poly(styrene sulfonate) film through solvent treatment. Polymer 45:8443–8450
21. Crispin X, Jakobsson FLE, Crispin A, Grim PCM, Andersson P, Volodin A et al (2006) The origin of the high conductivity of poly(3,4-ethylenedioxythiophene)-poly(styrenesulfonate) (PEDOT- PSS) plastic electrodes. Chem Mater 18:4354–4360
22. Park J, Jang JG, Kang K, Kim SH, Kwak J (2024) High thermoelectric performance in solution-processed semicrystalline PEDOT: PSS films by strong acid-base treatment: limitations and potential. Adv Sci 11:2308368
23. Patel SN, Glaudell AM, Kiefer D, Chabinyc ML (2016) Increasing the thermoelectric power factor of a semiconducting polymer by doping from the vapor phase. Acs Macro Lett 5:268–272
24. Patel SN, Glaudell AM, Peterson KA, Thomas EM, O'Hara KA, Lim E, Chabinyc ML (2017) Morphology controls the thermoelectric power factor of a doped semiconducting polymer. Sci Adv 3:e1700434
25. Mcculloch I, Heeney M, Bailey C, Genevicius K, Macdonald I, Shkunov M et al (2006) Liquid-crystalline semiconducting polymers with high charge-carrier mobility. Nat Mater 5:328–333
26. Kang K, Watanabe S, Broch K, Sepe A, Brown A, Nasrallah I et al (2016) 2D coherent charge transport in highly ordered conducting polymers doped by solid state diffusion. Nat Mater 15:896–902
27. Watanabe S, Ohno M, Yamashita Y, Terashige T, Okamoto H, Takeya J (2019) Validity of the Mott formula and the origin of thermopower in π-conjugated semicrystalline polymers. Phys Rev B 100:241201

28. Yamashita Y, Tsurumi J, Ohno M, Fujimoto R, Kumagai S, Kurosawa T et al (2019) Efficient molecular doping of polymeric semiconductors driven by anion exchange. Nature 572(2019):634–638
29. Chen C, Jacobs IE, Kang KH, Lin Y, Jellett C, Kang BS et al (2023) Observation of weak counterion size dependence of thermoelectric transport in ion exchange doped conducting polymers across a wide range of conductivities. Adv Energy Mater 13:2202797
30. Shi W, Deng TQ, Wong ZM, Wu G, Yang SW (2021) A molecular roadmap towards organic donor-acceptor complexes with high-performance thermoelectric response. npj Comput Mater 7:107
31. Ding JF, Yamanaka K, Hong SH, Chen GL, Wu WN, Lin JM et al (2024) Controlling the thermoelectric performance of doped naphthobisthiadiazole-based donor-acceptor conjugated polymers through backbone engineering. Adv Sci 11:2410046
32. Wang DY, Ding JM, Ma YQ, Xu CL, Li ZY, Zhang X et al (2024) Multi-heterojunctioned plastics with high thermoelectric figure of merit. Nature 632:528–535
33. deLeeuw DM, Simenon MMJ, Brown AR, Einerhand REF (1997) Stability of n-type doped conducting polymers and consequences for polymeric microelectronic devices. Synth Met 87:53–59
34. Pöhls JH, Johnson MB, White MA (2016) Origins of ultralow thermal conductivity in bulk [6,6]-phenyl-C-butyric acid methyl ester (PCBM). Phys Chem Chem Phys 18:1185–1190
35. Wei P, Oh JH, Dong GF, Bao ZN (2010) Use of a 1H-benzoimidazole derivative as an n-type dopant and to enable air-stable solution-processed n-channel organic thin-film transistors. J Am Chem Soc 132:8852–8853
36. Liu J, van der Zee B, Alessandri R, Sami S, Dong JJ, Nugraha MI, et al (2020) N-type organic thermoelectrics: demonstration of ZT>0.3. Nat Commun 11:5694
37. Schlitz RA, Brunetti FG, Glaudell AM, Miller PL, Brady MA, Takacs CJ et al (2014) Solubility-limited extrinsic n-type doping of a high electron mobility polymer for thermoelectric applications. Adv Mater 26:2825–2830
38. Wang Y, Nakano M, Michinobu T, Kiyota Y, Mori T, Takimiya K (2017) Naphthodithiophenediimide-benzobisthiadiazole-based polymers: versatile n-type materials for field-effect transistors and thermoelectric devices. Macromolecules 50:857–864
39. Wang SH, Sun HD, Erdmann T, Wang G, Fazzi D, Lappan U et al (2018) A chemically doped naphthalenediimide-bithiazole polymer for n-type organic thermoelectrics. Adv Mater 30:1801898
40. Wang Y, Takimiya K (2020) Naphthodithiophenediimide-bithiopheneimide copolymers for high-performance n-type organic thermoelectrics: significant impact of backbone orientation on conductivity and thermoelectric performance. Adv Mater 32:2002060
41. Liu J, Qiu L, Alessandri R, Qiu XK, Portale G, Dong JJ et al (2018) Enhancing molecular n-type doping of donor-acceptor copolymers by tailoring side chains. Adv Mater 30:1704630
42. Yang CY, Ding YF, Huang DZ, Wang J, Yao ZF, Huang CX et al (2020) A thermally activated and highly miscible dopant for n-type organic thermoelectrics. Nat Commun 11:3292
43. Tang HR, Liang YY, Liu CC, Hu ZC, Deng YF, Guo H et al (2022) A solution-processed n-type conducting polymer with ultrahigh conductivity. Nature 611:271–277
44. Ke ZF, Abtahi A, Hwang J, Chen K, Chaudhary J, Song I et al (2023) Highly conductive and solution-processable n-doped transparent organic conductor. J Am Chem Soc 145:3706–3715
45. Huang JD, Li QF, Wang QQ, Liu TF, Jeong SY, Paleti SHK et al (2025) Preventing benzoquinone-based catalyst aggregation enables the one-step synthesis of highly conductive poly(benzodifurandione) without post-reaction purification. Adv Mater 37:2502426
46. Glaudell AM, Cochran JE, Patel SN, Chabinyc ML (2015) Impact of the doping method on conductivity and thermopower in semiconducting polythiophenes. Adv Energy Mater 5:1401072
47. Wei Q, Mukaida M, Kirihara K, Naitoh Y, Ishida T (2016) Photoinduced Dedoping Of Conducting Polymers: An Approach To Precise Control Of The Carrier Concentration And Understanding Transport Properties. ACS Appl Mater Interfaces 8:2054–2060

Chapter 3
Anisotropic Characterization Protocols for Organic Thermoelectrics

Abstract Organic thermoelectric materials typically exhibit fibrous or layered microscopic structures, making it essential to account for the resulting physical anisotropy. Accurately characterizing these materials requires measuring the Seebeck coefficient, electrical conductivity, and thermal conductivity, at both in-plane and through-plane directions, to determine the figure of merit. However, because organic materials are primarily solution-processed into thin films rather than bulk ingots, complete directional characterization poses significant experimental challenges. This chapter details measurement protocols and technologies designed to overcome these barriers, enabling precise characterization across all directions.

Keywords Structural anisotropy · Measurement protocols · In-plane · Through-plane · Figure of merit

As discussed in previous sections, the performance of organic thermoelectric materials is intrinsically linked to their multi-scale morphology. A defining characteristic of polymer thin films is anisotropy, where electrical and thermal transport properties differ significantly depending on the direction of measurement—typically categorized as in-plane (parallel to the substrate) or through-plane (perpendicular to the substrate).

The origin of this anisotropy lies in the molecular alignment and hierarchical structure of the polymer chains. During solution-based processing, such as spin-coating or blade-coating, polymer backbones often align preferentially in the in-plane direction due to centrifugal forces or shear stress. Furthermore, the π-π stacking of conjugated systems often adopts a "face-on" or "edge-on" orientation relative to the substrate [1].

Because charge transport is much more efficient along the conjugated backbone and through π-π stacking than it is through disordered insulating regions (such as PSS domains), electrical conductivity (σ) is often orders of magnitude higher in the in-plane direction. Conversely, thermal conductivity (κ) tends to be lower in the through-plane direction because the layered structure acts as a barrier to phonon transport.

M. Mukaida and Q. Wei, *Stable Organic Thermoelectric Devices and Thermocells for Self-powered Sensor Applications*, SpringerBriefs in Materials, https://doi.org/10.1007/978-981-92-1038-1_3

3.1 Challenges in Directional Measurement

Standardizing characterization protocols for these anisotropic properties remains a significant challenge. For thin-film samples (typically sub-micrometer thickness), measuring through-plane properties is technically demanding due to the dominance of contact resistance and the difficulty of establishing a precise temperature gradient across a very small distance. In many reports, a "hybrid" approach is adopted—combining in-plane Seebeck coefficients and electrical conductivity with through-plane thermal conductivity (Fig. 3.1). However, it is critical to note that this methodology is physically inconsistent and will lead to an overestimation of the calculated figure of merit, *ZT*. For a rigorous assessment of performance, all parameters must be measured along the same crystallographic or geometric axis to account for the material's inherent structural anisotropy.

To address these challenges, our research group has focused on developing more rigorous and reliable characterization protocols for organic thermoelectric materials [2–4]. Recognizing that the common practice of mixing directional parameters leads to significant inaccuracies, we have established techniques that allow for the accurate and independent measurement of thermoelectric performance in both the in-plane and through-plane directions. By utilizing advanced methods—such as the lamellar technique for thermal diffusivity and specialized contact geometries for electrical transport—we aim to provide a self-consistent and comprehensive dataset for each material. This dual-direction accuracy is essential for bridging the gap between intrinsic material properties and practical module design, ensuring that our performance assessments reflect the true physical behavior of anisotropic polymer films.

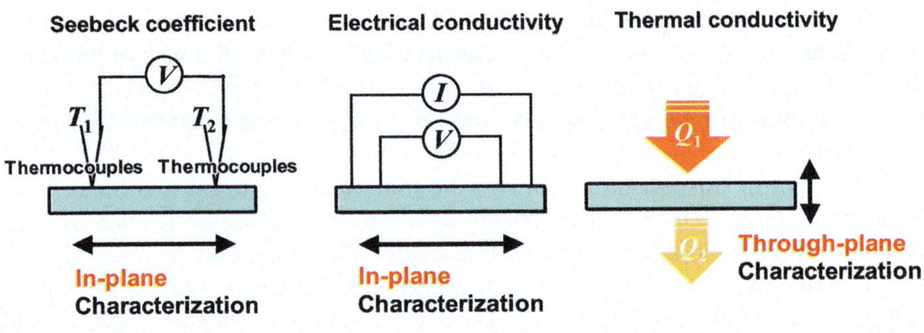

Fig. 3.1 Anisotropy evaluation image for Seebeck coefficient, electrical conductivity and thermal conductivity

3.2 Measuring the Through-Plane Seebeck Coefficient

The experimental setup for through-plane Seebeck coefficient (S) measurement is illustrated in Fig. 3.2. Because the Seebeck coefficient is determined by the voltage generated by a temperature gradient, establishing a precise temperature difference (ΔT) between two points is critical for accuracy. To achieve this, we utilize a device equipped with Peltier elements attached to the top and bottom of the film.

Due to the inherently low thermal conductivity of organic materials, it is possible to maintain a stable temperature difference of several degrees even across a film thickness of only a few tens of micrometers. The upper and lower Peltier elements are controlled to maintain an equal temperature interval from the center of the sample. For instance, increasing the top electrode by 2 °C while decreasing the bottom by 2 °C creates a total ΔT of 4 °C. By collecting data at a minimum of five distinct temperature points, we determine the relationship between the temperature gradient and the resulting thermovoltage. Measurement reliability is confirmed when the data exhibits a linear relationship that passes directly through the origin (zero point). To minimize error, each temperature point is recorded via a thermocouple, and the voltage is measured using the same type of wires from those thermocouples.

Using this specialized setup, we evaluated the through-plane Seebeck coefficient of a 100 μm-thick PEDOT:PSS organic film and a 1 mm-thick Nickel (Ni) foil. The through-plane S for PEDOT:PSS was measured at approximately 17 μV/K, which is identical to the in-plane value obtained via conventional methods. Similarly, the through-plane S for the isotropic Ni sample was measured at −19 μV/K, consistent with its known in-plane value [5]. In both cases, the relationship between the temperature difference and the voltage was linear and passed through the origin, indicating negligible measurement error. Crucially, our ability to accurately resolve the thermoelectric power in films at the micrometer scale demonstrates that this methodology is a robust and reliable tool for evaluating the through-plane Seebeck coefficient in organic film samples.

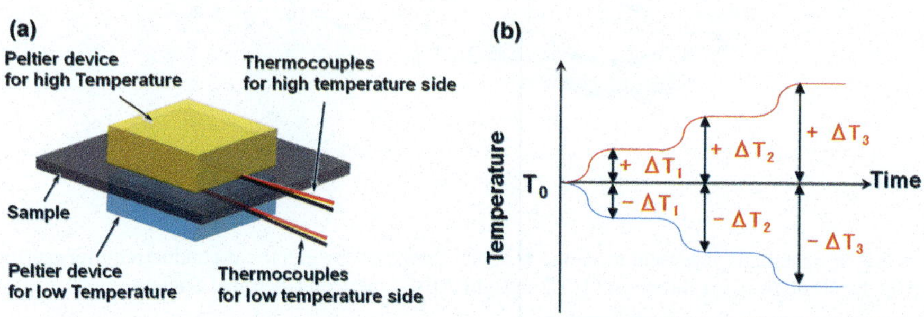

Fig. 3.2 Measurement method for S of through-plane direction: **a** schematic diagram and **b** diagram of the method of changing each temperature by the same amount over time

3.3 Through-Plane Electrical Conductivity Measurement

Measuring the electrical conductivity (σ) of thin films in the through-plane direction presents significant geometric challenges. Figure 3.3 provides a conceptual visualization of how we adapt the standard four-terminal method for this purpose.

3.3.1 Conceptual Evolution of the Measurement Geometry

The theoretical foundation of this technique relies on the assumption that a four-terminal measurement of through-plane conductivity becomes possible if the distance between the current and voltage probes is minimized toward zero.

In practice, physical probes always have a finite diameter and separation distance (*d*). However, by using the coaxial probe setup (Fig. 3.3), we can systematically vary the inter-electrode distance.

By measuring the resistance at several small, discrete distances and then extrapolating the results to $\boldsymbol{d = 0}$, we can mathematically simulate that "zero-distance" condition. This allows us to extract the A-component of the current—the part flowing strictly through the thickness—and achieve a measurement that is physically equivalent to a standard four-terminal evaluation, even in high-aspect-ratio thin films.

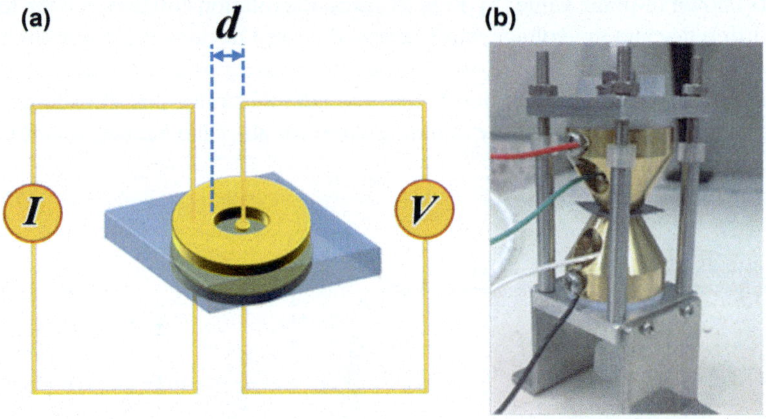

Fig. 3.3 **a** Schematic representation and **b** actual image of the electrical conductivity measuring device for through-plane direction [2]. Copyright 2014 American Chemical Society

3.3.2 Coaxial Four-Terminal Probe Setup

The practical implementation of this theory utilizes a coaxial four-terminal probe.

- **Electrode Design**: The probe consists of an outer electrode for current sourcing and an inner electrode for voltage measurement.
- **Materials**: All electrodes are constructed from gold-plated copper to ensure high conductivity and minimal contact resistance.
- **Variable Distance**: The distance between the electrodes (*d*) is made variable by utilizing inner electrodes of different diameters.
- **Insulation**: Precise electrical isolation between the inner and outer components is maintained via a polyimide dielectric layer.

3.3.3 Extrapolation and Validation

The relationship between the electrode distance (d) and the measured conductivity is fundamental to accurate data interpretation, as shown in Fig. 3.4 [3]. For high-conductivity isotropic materials like SUS304, the measured conductivity values vary significantly as a function of d. In these cases, extrapolating the data to a zero-distance limit is essential to eliminate geometric interference and reach the material's true through-plane conductivity.

In contrast, for anisotropic materials with high in-plane conductivity—such as Highly Oriented Pyrolytic Graphite (HOPG) or PEDOT:PSS—the influence of electrode distance is markedly reduced. This phenomenon can be attributed to a more uniform electric field distribution along the in-plane direction, which naturally suppresses lateral spreading and focuses the current path through the thickness.

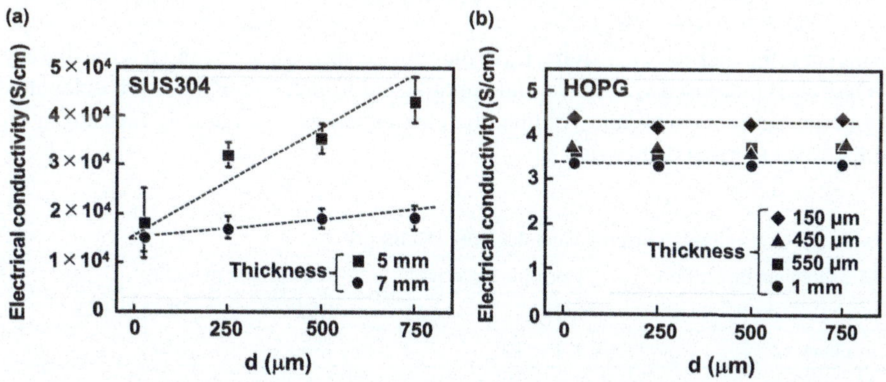

Fig. 3.4 Example of electrical conductivity measurement in the through-plane direction: **a** Measured electrical conductivity of SUS304 for different distances between the source probe and the sense probe and **b** the measured electrical conductivity of HOPG for different distances between the source probe and the sense probe [3]. Copyright 2016 Elsevier

Based on our extensive experimental experience with organic thermoelectric materials, we have found that while a small electrode spacing remains desirable to minimize potential error, these materials are generally less sensitive to the *d* parameter. This allows for a reliable and accurate assessment of through-plane conductivity even without exhaustive extrapolation, simplifying the characterization process for most organic thin films.

3.3.4 *Quantifying Anisotropy*

The robustness of this method is summarized in Table 3.1, which compares in-plane conductivity (measured via standard four-probe methods) with through-plane conductivity (measured via our coaxial probe). As shown, the isotropic nature of stainless steel (SUS304) is clearly reflected by the identical values in both directions. Conversely, the extreme anisotropy of HOPG and PEDOT:PSS—where conductivity varies by several orders of magnitude—is accurately captured. This ability to evaluate the through-plane component is essential for clearly distinguishing between isotropic and anisotropic transport regimes in organic electronics.

3.4 Characterizing Thermal Conductivity

To fully evaluate the figure of merit (*ZT*), accurate determination of thermal conductivity (κ) is essential. Thermal conductivity is typically derived using the following relationship:

$$\kappa = \alpha \times C_p \times \rho$$

where α is the thermal diffusivity, C_p is the specific heat capacity, and ρ is the density. While C_p is measured via differential scanning calorimetry (DSC) and ρ is calculated from the sample's mass and volume, the challenge lies in accurately measuring the directional thermal diffusivity, α.

Table 3.1 Anisotropy of electrical conductivity (S/cm) [2]

Measurement direction	Isotropic material	Anisotropic materials	
	SUS	HOPG	PEDOT:PSS
In-plane	1.3×10^4	430	820
Through-plane	1.6×10^4	3.4	36

3.4.1 The Flash and Lamellar Methods

Figure 3.5 illustrates the measurement of α using the flash method (also known as the pulse heating method). In this setup, the front surface of a sample is subjected to a high-energy pulse from a xenon lamp or laser. The resulting temperature rise is monitored on the back surface.

By analyzing the temperature–time transient, the through-plane thermal diffusivity can be determined. This method relies on a one-dimensional unsteady heat transfer model designed to minimize heat loss in other directions by completing the measurement in a very short time interval. For a sample of thickness *d*, if T_{max} is the maximum temperature rise and $t_{1/2}$ is the "half-time" (the time required to reach half of T_{max}), the thermal diffusivity is calculated as:

$$\alpha = 0.1388\frac{d^2}{t_{1/2}}$$

Measuring thermal diffusivity in the in-plane direction of thin films is traditionally difficult with standard flash equipment. To overcome this, we utilize the lamellar method, which adapts the flash technique through specific sample processing (Fig. 3.6).

1 **Preparation**: The thin-film sample is cut into uniform strips (a).
2 **Assembly**: These strips are tightly rolled or stacked into a bulk-like cylindrical specimen (b).
3 **Orientation**: The processed sample is rotated 90 degrees (c), effectively orienting the original "in-plane" surface of the film to face the pulse source and the detector.

By measuring this reconstructed sample, the flash device can accurately probe the thermal transport that originally occurred along the film's surfacc.

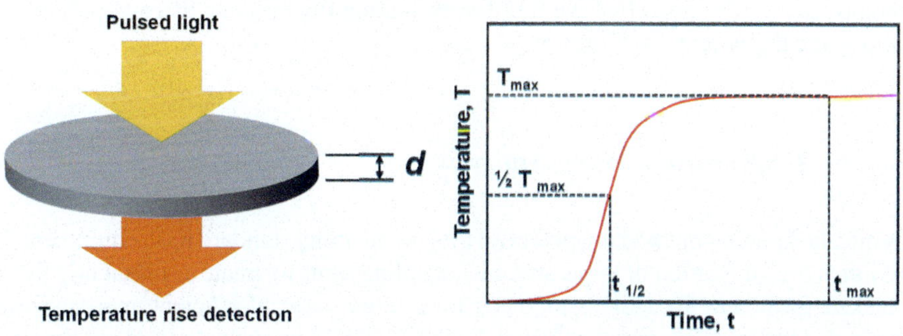

Fig. 3.5 Schematic diagram and measurement image of pulse heating method (flash method)

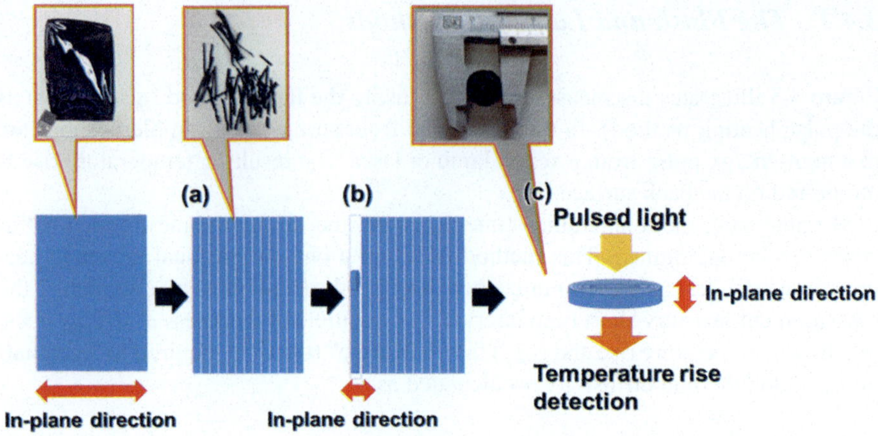

Fig. 3.6 Schematic diagram of the lamellar method. **a** cut into uniform strips, **b** rolled into a bulk-like cylindrical specimen and **c** the sample is rotated 90 degrees [2] Copyright 2014 American Chemical Society

3.4.2 Experimental Results and Analysis

Figure 3.7 displays the thermal measurement data for a 100 μm-thick PEDOT:PSS sheet obtained via the flash method [5, 6]. The temperature rise curves for the through-plane direction (a) and the in-plane direction (b) show distinct characteristics. Notably, the in-plane data (b) exhibits higher noise levels compared to the through-plane measurements. This is attributed to the physical nature of the lamellar sample; the stacked or "wrapped" edges create a slightly uneven surface on both the irradiated front and the monitored back. This unevenness causes a portion of the light to scatter. However, despite this increased noise, a clear and reproducible temperature rise trend is obtained. This demonstrates that the lamellar-adapted flash method is a robust and effective tool for evaluating the in-plane thermal properties of organic thermoelectric films.

3.4.3 Temperature Wave Analysis

While the lamellar method is a powerful tool for adapting standard flash equipment, it requires the fabrication of large-area free-standing films to create a sufficiently thick stacked specimen. To address the need for a more material-efficient approach, we have also utilized Temperature Wave Analysis (TWA) to evaluate in-plane thermal diffusivity [4]. As illustrated in the methodology established in our 2016 study, this technique utilizes a micro-heater and a temperature sensor placed in direct contact with a single free-standing film (Fig. 3.8).

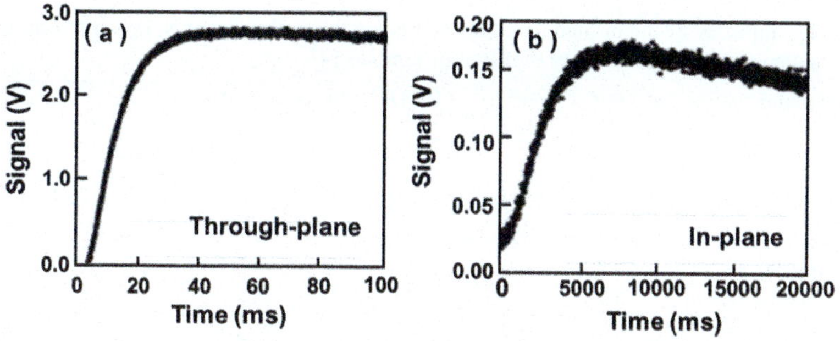

Fig. 3.7 Temperature rise curves of the PEDOT:PSS film in the **a** through-plane and **b** in-plane directions, which using the lamellar method [2]. Copyright 2014 American Chemical Society

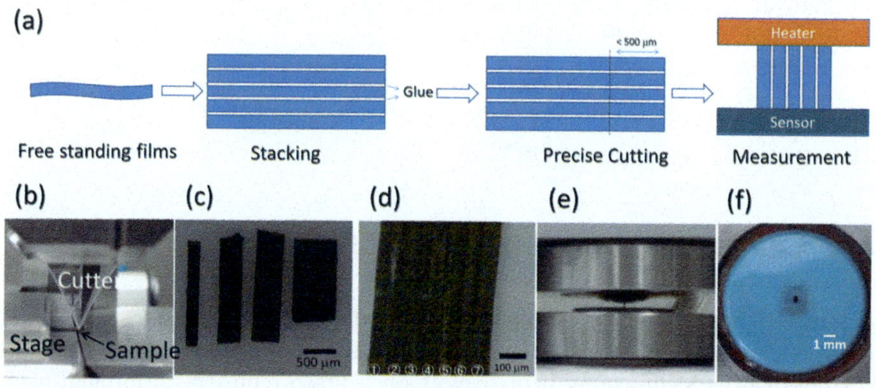

Fig. 3.8 **a** Schematic of the in-plane thermal diffusivity measurement used in this study. **b** A photograph of the film sample on the precise cutter. The microscope images of the prepared samples with **c** top view and **d** side view. **e** A photograph of the film sample standing between heater and sensor. [4]. Copyright 2016 American Institute of Physics

The primary advantage of the TWA method is its minimal material requirement; it can accurately resolve the in-plane thermal properties of a single film with a thickness of only 10–100 μm, eliminating the need for the complex stacking and rolling processes required by the lamellar-flash method. By analyzing the phase shift of a temperature wave as it propagates through the film, we have successfully quantified the thermal anisotropy of various organic systems (Table 3.2).

Most samples in our study exhibited anisotropic thermal conductivity; however, PDMS (polydimethylsiloxane) was found to be nearly isotropic. This behavior can be explained by the relationship between chemical structure and processing conditions. It is well-established that heat transport is significantly more efficient through intrachain covalent bonding than through interchain van der Waals interactions. During typical film fabrication, most polymer chains adopt a "preferred" orientation, leading to anisotropic heat conduction.

Table 3.2 Summary of the in-plane and through-plane thermal diffusivity, density, heat capacity, and calculated thermal conductivity of different samples [4]

Sample name	α // (m^2/s)	α⊥ (m^2/s)	Cp (J/gK)	Density (g/cm^3)	Λ// (W/mK)	Λ⊥ (W/mK)	Anisotropic ratio
Poly(dimethylsiloxane) (PDMS)	1.1×10^{-7}	1.0×10^{-7}	1	1	0	0	~1
Polyethylene terephthalate (PET)	3.0×10^{-7}	1.0×10^{-7}	1	1	0	0	3.0
Nafion N-1154	2.1×10^{-7}	9.3×10^{-8}	1	2	0	0	2.3
Poly(3-hexylthiophene) (P3HT)	2.3×10^{-7}	1.0×10^{-7}	2	1	0	0	2.3
PEDOT:PSS (1.7 S/cm)	4.0×10^{-7}	1.1×10^{-7}	1	2	1	0	3.6
PEDOT:PSS (810 S/cm)	5.2×10^{-7}	1.1×10^{-7}	1	2	1	0	4.7
Carbon paper (TGP-H-120)	6.9×10^{-5}	4.4×10^{-6}	1	0	22	1	16

3.5 Comparative Anisotropy Analysis

Figure 3.9 provides a comprehensive summary of the directional anisotropy observed in the Seebeck coefficient (S), electrical conductivity (σ), and thermal conductivity (κ) for PEDOT:PSS thin films. In these plots, open circles represent the through-plane direction, while filled circles denote the in-plane direction.

- **Seebeck Coefficient (S)**: No significant anisotropy is observed. The S values remain largely consistent regardless of the measurement orientation, suggesting that the carrier energy distribution is relatively uniform across the polymer domains.

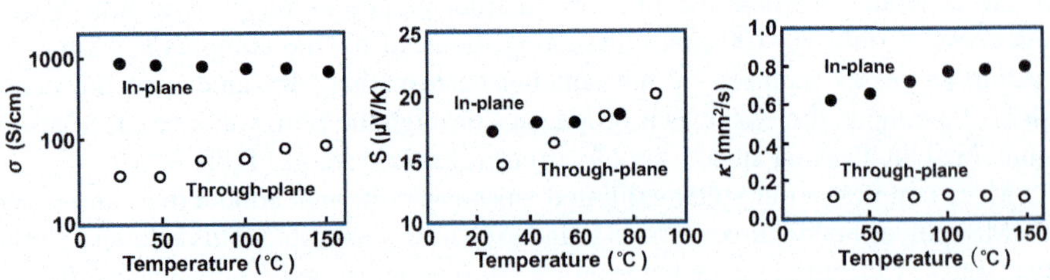

Fig. 3.9 Anisotropy of σ, S and κ properties of PEDOT:PSS [2]. Copyright 2014 American Chemical Society

- **Electrical Conductivity** (**σ**): Substantial anisotropy is present, with in-plane conductivity exceeding through-plane values by several orders of magnitude. This is a direct consequence of preferential chain alignment and π-π stacking along the substrate.
- **Thermal Conductivity** (κ): Significant anisotropy is also observed here. As expected, κ is lower in the through-plane direction, where the layered lamellar structure of the polymer acts as a barrier to phonon transport.

The comprehensive characterization of thermoelectric properties in both directions provides several critical insights for the rational design of organic energy harvesters. First, accurately measuring both in-plane and through-plane parameters is the only way to avoid the common pitfall of overestimating the figure of merit ZT. Relying on "hybrid" values—such as combining high in-plane conductivity with low through-plane thermal conductivity—creates a physically inconsistent and artificially inflated ZT that does not represent the actual performance of the material. Second, our analysis confirms that the in-plane ZT is typically an order of magnitude larger than the through-plane ZT. While the Seebeck coefficient remains largely isotropic, the superior charge transport along the oriented polymer backbones leads to in-plane electrical conductivities that far outweigh the benefit of the lower through-plane thermal conductivity. Finally, these stark differences in directional performance dictate that structural anisotropy must be a primary consideration in device engineering. The orientation of the polymer chains relative to the temperature gradient will fundamentally affect module architecture and efficiency, a topic that we will explore in detail in the following sections on module design.

References

1. Marina A, Spoelstra S, Zongtag HA, Wemmers AK (2021) An estimation of the European industrial heat pump market potential. Renew Sustain Energy Rev 139:110545
2. Wei Q, Mukaida M, Kirihara K, Ishida T (2014) Experimental studies on the anisotropic thermoelectric properties of conducting polymer films. Acs Macro Lett 3:948–952
3. Wei Q, Suga H, Ikeda I, Mukaida M, Kirihara K, Naitoh Y, Ishida T (2016) An accurate method to determine the through-plane electrical conductivity and to study transport properties in film samples. Org Electron 38:264–270
4. Wei Q, Uehara C, Mukaida M, Kirihara K, Ishida T (2016) Measurement of in-plane thermal conductivity in polymer films. AIP Adv 6:045315
5. Burkov AT, Heinrich A, Konstantinov PP, Nakama T, Yagasaki K (2001) Experimental set-up for thermopower and resistivity measurements at 100–1300 K. Meas Sci Technol 12:264–272
6. Weathers A, Khan ZU, Brooke R, Evans D, Pettes MT, Andreasen JW et al (2015) Significant electronic thermal transport in the conducting polymer Poly(3,4-ethylenedioxythiophene). Adv Mater 27:2101–2106

Chapter 4
Device and Module Design

Abstract For thermoelectric technologies to transition from laboratory research to practical application, advanced material performance must be coupled with strategic module design and system integration. Because organic thermoelectric materials are predominantly fabricated as thin films, unique engineering approaches are required to modularize them effectively. This chapter explores design methodologies for maximizing power output per unit area and volume. Detailed guidelines are provided for minimizing this contact resistance, mitigating the reduction of the effective temperature difference, and constructing all-organic laminated modules. Furthermore, practical demonstrations illustrate how this harvested power can simultaneously drive multiple sensors and Bluetooth communication units, as well as recharge commercially available lithium-ion secondary batteries.

Keywords Module design · Lamination · Interfacial resistance · DC-DC converters · Battery charging

4.1 Thin-Film Thermoelectric Devices

As we transition from the fundamental characterization of anisotropic materials to their practical implementation, it becomes clear that material performance is only one half of the equation. To translate these intrinsic properties into usable energy, we must shift our focus to Module Design and System Integration.

In practical applications, the electrical output of a single thermoelectric element is insufficient to power modern electronics. Therefore, individual p-type and n-type units are electrically connected in series and thermally in parallel to form a thermoelectric module.

M. Mukaida and Q. Wei, *Stable Organic Thermoelectric Devices and Thermocells for Self-powered Sensor Applications*, SpringerBriefs in Materials, https://doi.org/10.1007/978-981-92-1038-1_4

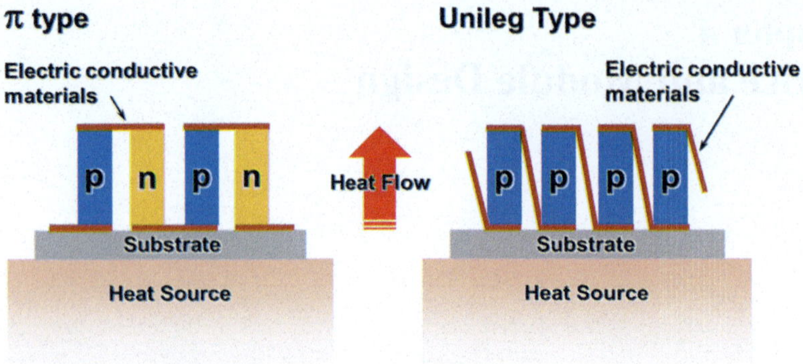

Fig. 4.1 Schematic diagrams for basic module designs of π type and unileg type

As illustrated in Fig. 4.1, there are two primary architectures for connecting organic thermoelectric materials:

- **π-type Configuration**: This is the standard architecture where alternating p-type and n-type legs are connected. This design is highly advantageous because the high-temperature and low-temperature junctions are physically separated, maintaining the effective temperature difference.
- **Unileg Configuration**: Historically, because stable n-type organic semiconductors were difficult to achieve in the organic thermoelectric category, much of the early work in the field focused on "unileg" modules. These consist solely of p-type legs connected by metallic interconnects. While easier to fabricate when n-type materials are unavailable, they generally suffer from lower efficiency due to the thermal connection between the hot and cold areas.

Unlike traditional inorganic thermoelectrics, which are typically processed as rigid bulk ingots, organic materials are solution-processable and generally used as thin films. This necessitates a fundamental rethink of module geometry. While the inorganic industry relies on a "pellet" model (e.g., bismuth telluride), organic materials require us to effectively stack, fold, or print films.

Since organic thermoelectric raw materials are typically liquid-based, they are naturally suited for high-throughput, large-area manufacturing. Techniques such as screen printing, roll-to-roll processing, and inkjet printing are the primary methods considered for scalability. Looking forward, these liquid precursors could even be adapted for coating fabrics or functionalizing individual threads, though fabricating stable, pure organic thermoelectric fibers remains a challenge.

Our group's early efforts focused on a straightforward but practical starting point: printing organic materials directly onto paper substrates using a unileg structure. While a unileg design (using only p-type legs) is not the most "elegant" or efficient from a performance perspective, it is a highly accessible "smart-start" strategy. It allowed us to rapidly prototype using existing printing technologies before stable n-type polymers were widely available.

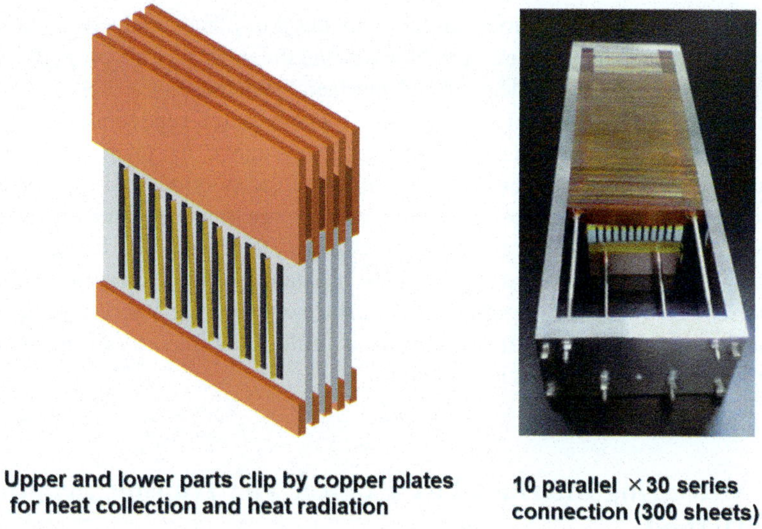

Fig. 4.2 Schematic diagram and photograph of screen-printed paper module sandwiched between copper plates [1]. Copyright 2014 Royal Society of Chemistry

As shown in Fig. 4.2, we developed a method where a PEDOT:PSS solution (containing 3% EG as a dopant) is applied as an ink onto a paper substrate through a metal mask [1]. After the solvent evaporates, several rows of the organic material are left on the paper. To overcome the low voltage of a single column, multiple columns are connected in series using conductive silver paste.

A single paper element, such as the one shown in the photograph containing 11 columns of PEDOT:PSS, still produces limited output. To generate meaningful power, these individual paper sheets must be integrated into a larger module.

- Series Configuration: Increases the total output voltage.
- Parallel Configuration: Reduces the overall internal electrical resistance.

In a benchmark study, we constructed a module consisting of 300 sheets (organized as 10 in parallel and 30 in series). When one end was placed in contact with a 180 °C heat source and the other cooled by forced air, the module maintained a temperature difference ΔT of 100 K. This setup generated 50 μW of power. With a module volume of 10 cm × 40 cm × 7 cm, the resulting power density was approximately 0.12 μW/cm^2.

Despite the successful demonstration of the paper-based module, this study yielded several unexpected and "strange" results that provided deep insights into the physics of organic device integration. While the measured open-circuit voltage (V_{oc}) aligned closely with the values predicted by our in-plane Seebeck coefficient characterization, the internal resistance of the total module was significantly higher than the theoretical expectations based on standard in-plane conductivity measurements. This discrepancy forced us to look beyond the intrinsic bulk properties of the polymer and identify the critical role of interfacial resistance.

Unlike inorganic thermoelectric modules, where metal–semiconductor contacts are well-optimized and often exhibit negligible resistance, the interfaces in organic devices—specifically between the solution-processed polymer and the conductive interconnects—introduce substantial barriers to charge transport. These results highlighted that in the case of organic thermoelectrics, the loss at the interface can be a dominant factor, often dwarfing the bulk resistance. This realization shifted our research focus toward the importance of interface engineering and contact optimization, as maximizing the ZT of the material is futile if the majority of the generated power is dissipated at the device junctions.

4.2 Measuring Interfacial Resistance: The Transfer Length Method

To mitigate the high internal resistance observed in our initial prototypes, we focused on screening interfacial materials to minimize contact resistance (R_c).

We evaluated these interfacial losses using the Transfer Length Method (TLM), as illustrated in Fig. 4.3. in this setup, a conductive material (such as a metal foil) is placed on top of a PEDOT:PSS film at varying distances (L). Probes are placed on the electrodes to measure the total electrical resistance (R).

The total measured resistance R is the sum of the bulk resistance of the PEDOT:PSS (R_s) and the two contact resistances (R_c) occurring at the interfaces:

$$R = R_s + 2R_c$$

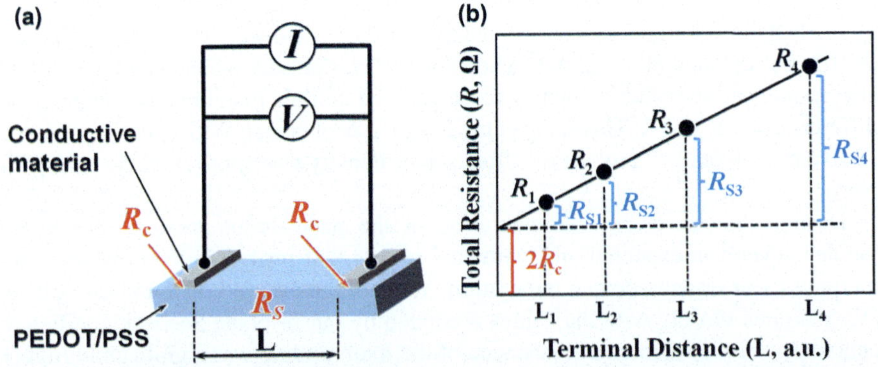

Fig. 4.3 **a** Schematic setup image and **b** relationship between total resistance and terminal distance by using of transfer length method for evaluating contact electrical resistance

Table 4.1 Contact resistivity between PEDOT:PSS and different materials

Materials	Al	Ni	Cu	Carbon	Ag	Au/Ni	Au	Au(Evap)	Pt
Contact resistivity, R_c ($\mu\Omega$ cm^2)	2.64×10^7	1.77×10^5	9.50×10^4	4.03×10^4	2.55×10^4	2.33×10^4	2.08×10^4	1.79×10^4	1.27×10^4

By plotting R against the distance L, we obtain a linear relationship. The slope represents the resistivity of the polymer, while the intercept at $L = 0$ provides the value for 2 R_c. This allows us to isolate the resistance purely attributed to the interface.

4.3 Comparative Analysis of Contact Materials

Table 4.1 summarizes the contact electrical resistance for various metals and coatings layered on PEDOT:PSS. Our findings indicate that interfacial resistance can vary by several orders of magnitude depending on the chosen material. Interestingly, while Gold (Au, 5.47 eV) and Nickel (Ni, 5.22 eV) have similar work functions, their contact resistances differ drastically. This discrepancy is attributed to the surface state of the metals; Al and Ni naturally form passive oxide layers that act as electrical insulators, hindering carrier injection from the organic material. In contrast, noble metals like Au and Pt maintain a clean, conductive surface. From a manufacturing perspective, using pure precious metals for module interconnects is cost-prohibitive. However, we demonstrated that a practical compromise is to use a thin Au coating on a Ni substrate (Au/Ni). As shown in Table 4.1, the contact resistance of Au/Ni (2.3×10^4 $\mu\Omega$cm^2) is nearly identical to that of vacuum-deposited gold (Au(Evap)) or pure gold foil. This finding provides a cost-effective pathway for module fabrication, ensuring high conductivity and stable interfaces without the need for expensive bulk precious metals.

4.4 The Lamination Strategy: Uni-Leg Modules with Ni Foil

To improve the power density and compactness of organic thermoelectric systems, we transitioned from bulky paper-based architectures to a high-density lamination process. The core of this strategy involves tightly sandwiching a thin insulating film between a p-type organic layer and a metallic interconnect.

4.4.1 Material Synergy: PEDOT:PSS and Nickel

In this "uni-leg" configuration, we utilized Nickel (Ni) foil not only as an electrical interconnect but also as a functional thermoelectric partner. Nickel is uniquely suited for this role because it possesses one of the highest negative Seebeck coefficients among common metals, effectively serving as an "n-type" counterpart. While the thermal conductivity of PEDOT:PSS is nearly two orders of magnitude lower than that of Ni, its electrical conductivity is also roughly 1/100 that of the metal. This symmetry results in thermoelectric figures of merit (Z) that are remarkably well-matched, creating a balanced and efficient series connection.

4.4.2 Fabrication via Precision Punching

To ensure high dimensional accuracy, we employed a precision punching method to prepare the various layers:

- **PEDOT:PSS (50 μm)**: Cast from a 3% EG solution and punched into 22 mm × 22 mm.
- **Nickel Foil (5 μm)**: High-purity (>99%) thin films punched into 22 mm × 22 mm.
- **Insulating Film (5 μm)**: Polyimide (PI) sheets punched to a slightly smaller size 20 mm × 22 mm to allow the PEDOT:PSS and Ni to make contact at the edges.

However, as highlighted in our earlier discussion on interfacial resistance, simply stacking these layers resulted in an abnormally large internal resistance. This barrier was so significant that it initially made power generation impossible, regardless of the voltage produced. In a thermoelectric module, achieving a high voltage is futile if the internal resistance prevents current from flowing to the load.

To clarify and resolve this issue, we focused on the contact points between the PEDOT:PSS and the Ni foil [2]. As shown in Fig. 4.4, we implemented two critical refinements:

1. **Selective Metallization:** Gold (Au, 20 nm thick) was vapor-deposited onto the Ni foil strictly at the contact points to eliminate the resistive oxide layer.
2. **Mechanical Compression:** After assembly, the laminated stack was compressed at approximately 3 MPa. This high-pressure treatment eliminates micro-scale unevenness at the interfaces, maximizing the effective contact area.

These optimizations led to a significant breakthrough. As shown in the figure, a compact module laminating 50 PEDOT/PSS films (and 49 Ni films) achieves an output of 37 μW at a temperature difference of 50 K. This corresponds to a power density of 24 $\mu W/cm^2$. This represents a significant improvement over the approximately 0.12 $\mu W/cm^2$ of the sheet-type module introduced in the previous section. Despite these gains, a practical limitation remains: in this specific prototype, the temperature difference was maintained via forced air cooling. Since the energy

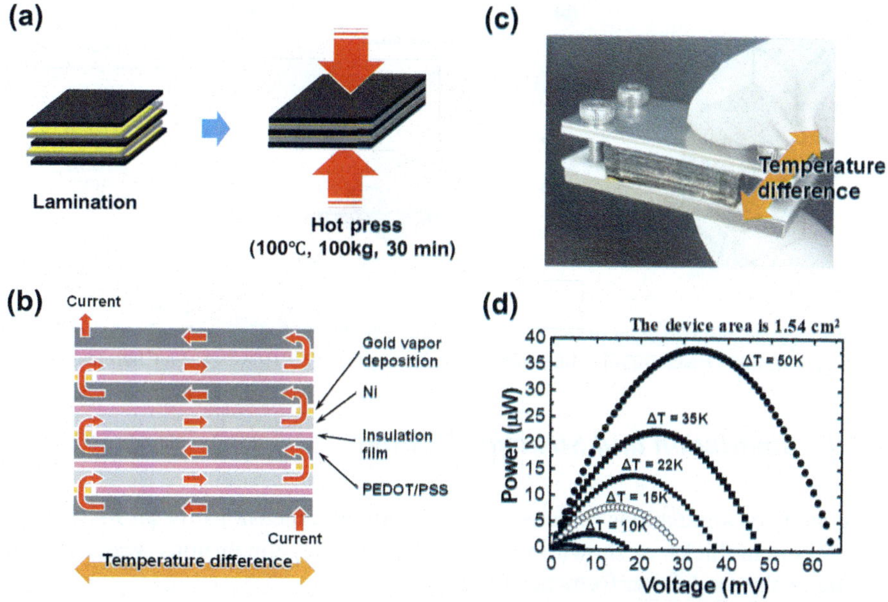

Fig. 4.4 **a** Schematic of the fabrication, **b** image of electrical current, **c** photograph and **d** power output diagram of the laminated module [2]. Copyright 2017 Elsevier

consumed by the cooling system currently outweighs the power generated, this setup cannot yet be classified as a truly "self-sustained" or energy-saving power source. Moving forward, the focus must shift toward passive cooling designs and harvesting heat from natural ambient gradients.

To further optimize the laminated module for real-world applications, we addressed the critical challenge of maintaining a temperature difference without active, energy-consuming cooling. The primary bottleneck in our previous "uni-leg" design was the high thermal conductivity of the Nickel (Ni) foil, which acted as a thermal bridge, rapidly equalizing the temperature between the hot and cold ends.

4.4.3 Geometric Optimization: The I-Shaped Ni Interconnect

Since the thermal conductivity of metals is more than two orders of magnitude higher than that of organic polymers, we sought a geometric solution to increase the thermal resistance of the Ni component without excessively increasing its electrical resistance. By strategically thinning or narrowing the metal interconnect, we can hinder heat bypass while leveraging the metal's superior electrical conductivity.

4.5 All-Polymer Laminated Modules: The PBFDO/PEDOT:PSS System

Building on the lamination strategy used in our Ni/polymer modules, we transitioned to a fully organic, all-polymer laminated module by substituting the metal foils with high-performance n-type polymers in Fig. 4.7 [4]. As reported in our recent study, the emergence of poly(benzodifurandione) (PBFDO)—which exhibits a breakthrough n-type conductivity exceeding 2000 S/cm—has finally made this metal-free architecture a reality.

The design principle remains the same as our previous Ni-based modules: p-type and n-type layers are alternatingly stacked with thin insulating films to create a high-density vertical device. Replacing Ni with PBFDO offers several unique advantages.

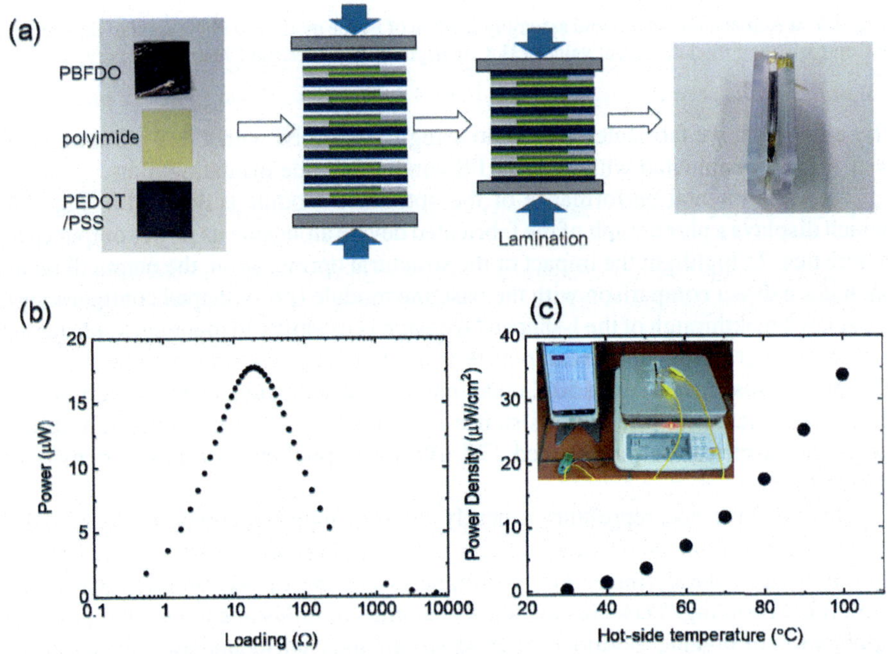

Fig. 4.7 **a** Schematic of the thermal lamination process used to fabricate the all-polymer TE module. PEDOT:PSS and PBFDO films were cut into 22 × 22 mm^2 squares, whereas polyimide films were trimmed to 22 × 20 mm^2. The layers were aligned and laminated at 50 °C under pressure. The photographs of the freestanding polymer films and the assembled module are also shown. **b** Power output of the device comprising 50 layers (25 p–n pairs connected in parallel) of PEDOT:PSS and PBFDO. **c** Power density of the device as a function of hot-side temperature. Inset: Demonstration of wireless sensor operation with data transmission to a smartphone via Bluetooth [4]. Copyright 2026 American Institute of Physics

The all-polymer module demonstrated exceptional environmental durability, maintaining consistent interfacial resistance and power output for more than one year in ambient conditions—a significant hurdle for previous n-type organic electronics.

Furthermore, this metal-free construction introduces a new paradigm of sustainable electronics. Because the entire module consists of organic polymers and combustible substrates, it can be disposed of via clean incineration at the end of its life cycle. This "from design to disposal" approach ensures that the large-scale deployment of self-powered IoT sensors does not contribute to the growing global burden of electronic waste.

This work confirms that by combining anisotropic structural design with high-conductivity n-type polymers, all-polymer modules can now rival hybrid systems in performance while offering superior stability and environmental compatibility.

Table 4.2 provides a comprehensive comparison of the PEDOT-based modules developed throughout our research, tracing the evolution from initial proofs-of-concept to high-efficiency laminated systems. We began with paper-based "unileg" modules, which prioritized manufacturing ease via high-throughput screen printing but suffered from low power densities due to bulky substrates and significant interfacial losses. By transitioning to a compact lamination architecture and addressing contact resistance through selective gold-coating and high-pressure compression (3 MPa), we achieved a massive leap in performance. The implementation of I-shaped Ni interconnects further optimized the balance between electrical and thermal resistance, enabling the module to reach 40 μW/cm^2 under natural cooling conditions. Finally, our most recent advancement—the all-polymer laminated module—replaces metal foils with high-performance n-type PBFDO, resulting in a metal-free, environmentally sustainable device that maintains high power density and exceptional long-term stability. This progression demonstrates that while intrinsic material ZT is a vital starting point, the ultimate success of organic thermoelectrics relies on the strategic engineering of device geometry and interfacial transport.

4.6 Demonstrations and Applications

To transition from a standalone thermoelectric module to a practical power source for electronic sensors, the integration of a boost circuit is essential. Organic thermoelectric modules typically generate low output voltages—often in the range of tens to hundreds of millivolts—whereas most microelectronic components require a stable supply of 1.8 V–3.3 V to operate.

We evaluated the efficiency of a commercially available DC/DC boost converter (Matrix) as a function of both input voltage and internal resistance (Fig. 4.8). Our investigation revealed a critical design guideline: minimizing internal resistance is key to circuit efficiency. Lower internal resistance allows the converter to "kick in" and operate efficiently at much lower input voltages. For practical IoT applications, a target efficiency of 60% is desirable; our data indicates that if the internal resistance

Table 4.2 Composition table for laminated modules [1–4]

Type	Sheet	Lamination	Heat flow optimized	p-n (all organic)
Exterior				
Module size (cm) Width × Depth × Height	10 × 40 × 7	2.2 × 0.7 × 2.2	2.2 × 0.4 × 2.2	2.2 × 0.24 × 2.2
Heat source contact area (cm^2)	400	1.54	0.9	0.53
Maximum power (μW)	50	37	36	18
Power density ($\mu W/cm^2$)	0.12	24	40	34
Number of element	300	100	50	50 (25p + 25n)

is maintained near 10 Ω, an input voltage of only 50 mV is sufficient to trigger the boost converter.

To meet these electrical requirements, we developed an optimized module with an overall size of 22 mm × 22 mm × 6 mm and a weight of approximately 1 g [5]. When subjected to a hot-end temperature of 100 °C, the module achieves the following:

- Power Output: Exceeds 90 μW
- Power Density: 78 $\mu W/cm^2$
- Internal Resistance: Maintained at a stable 11 Ω.

We successfully demonstrated this module powering multiple sensors and a Bluetooth communication unit under natural cooling conditions. The system transmitted data to a smartphone over a distance of 20 m in open space and maintained reliable communication even in "noisy" industrial factory environments (10–20 m distance). Long-term durability testing confirmed uninterrupted operation for 2 months, with frequent evaluations suggesting that the module's performance can be maintained for years, proving its viability as a long-term, self-sustained power source.

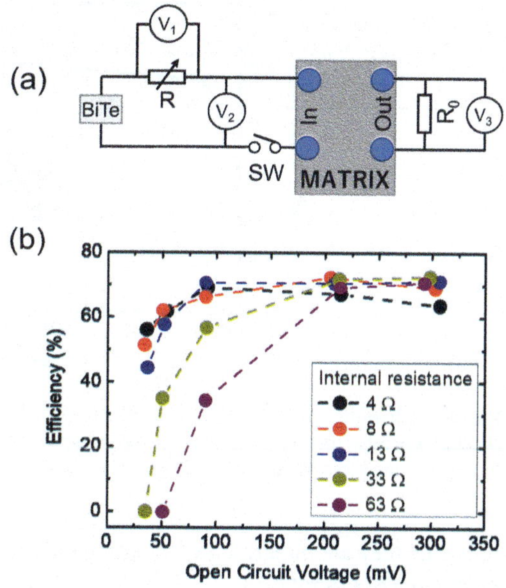

Fig. 4.8 **a** Block diagram of the configuration utilized to measure the conversion efficiency of the DC-to-DC converter (DC: direct current). The open-circuit voltage is controlled by a BiTe-alloy-based thermoelectric generator (BiTe). **b** DC-to-DC converter efficiency as a function of the open-circuit voltage of thermoelectric modules characterized by different values for the internal resistance [5]. Copyright 2023 Elsevier

4.7 Recharging Lithium-Ion Batteries

While thermoelectric modules are effective standalone power sources, integrating them directly into existing electronics often requires a total redesign of the device's power management system and manufacturing lines. While such shifts may be inevitable in the long term, a more immediate and practical strategy to promote thermoelectric technology is to use it as a complementary power solution—specifically for battery recharging.

Most IoT sensors currently rely on rechargeable Lithium-ion (Li-ion) secondary batteries. These batteries are typically charged using AC adapters that step down 100–220 V power to a few volts, a process that wastes a significant amount of energy as heat. By harvesting waste heat to charge these batteries directly, we can bypass the inefficiencies of grid-based charging and contribute to a meaningful reduction in global carbon emissions.

As a proof of concept, we tested the system using the smallest capacity Li-ion battery commercially available (Nichicon SLB3070LR35; rated voltage: 2.4 V; capacity: 0.35 mAh). With the hot-end temperature set to 100 °C, the battery achieved a full charge within 36 h as shown in Fig. 4.9.

Organic thermoelectric materials currently fall short of state-of-the-art inorganic semiconductors when measured strictly by conventional ZT benchmarks. However, as we have demonstrated, it is essential to recognize that ZT alone is not the decisive metric for the future of self-powered IoT. The power requirements of modern sensors and communication modules have plummeted into the microwatt and nanowatt regimes. In this landscape, organic materials already deliver sufficient power density,

Chapter 5
Thermocells

Abstract Thermocells, or thermogalvanic cells, offer an alternative electrochemical pathway for energy harvesting, generating electricity through temperature-dependent redox reactions. This mechanism naturally yields exceptionally large Seebeck coefficients, creating a significant potential difference when a thermal gradient is applied across the cell's electrodes. This chapter details the development of both single-cell architectures and multi-leg module configurations. This chapter highlights recent advancements in quasi-solid-state thermocells, detailing how the integration of advanced gel electrolytes ensures robust mechanical stability without sacrificing electrochemical performance.

Keywords Thermocells · Redox reactions · Gel electrolytes · Electrode

5.1 Liquid-State Thermoelectric Energy Harvesting

Throughout the preceding sections, we explored the immense potential of solid-state organic thermoelectric materials, culminating in the development of highly engineered, all-polymer modules. These devices rely fundamentally on the diffusion of electronic charge carriers—electrons and holes—across a temperature gradient. However, electronic transport is not the only mechanism capable of harvesting ambient heat. By shifting our focus from solid polymer networks to liquid electrolyte solutions, we open the door to an entirely different paradigm of energy conversion: the thermocell. Unlike their solid-state counterparts, thermocells (or thermogalvanic cells) generate electricity through temperature-dependent redox reactions and the migration of ions.

This unique electrochemical mechanism yields exceptionally high Seebeck coefficients—often an order of magnitude larger than those found in conjugated polymers—positioning liquid-state systems as a powerful parallel technology for capturing low-grade thermal energy and powering the next generation of IoT devices.

M. Mukaida and Q. Wei, *Stable Organic Thermoelectric Devices and Thermocells for Self-powered Sensor Applications*,
SpringerBriefs in Materials, https://doi.org/10.1007/978-981-92-1038-1_5

directly powering the majority of modern microelectronics, which typically require voltages in the range of 1.8 V–3.3 V.

5.6.2 *Single-Cell Systems with DC-DC Converters*

For single-cell applications, we have demonstrated successful energy harvesting by pairing the electrochemical cells with specific power management circuits designed for low-voltage startup: A compact 1.5 cm^3 device could successfully engaged an Asahi Kasei AP4473 converter, demonstrating the ability to power indicator LEDs solely from metabolic heat. A larger 2.5 $\times$ 2.5 $\times$ 1.5 cm^3 device can be paired with a Matrix Mercury converter. This system enabled the continuous, battery-free operation of a Bluetooth Low Energy (BLE) sensor node using solar-thermal energy [23].

5.6.3 *Series-Connected High-Voltage Modules*

To eliminate the need for voltage-boosting electronics and their associated efficiency losses, we developed a series-connected 35-leg module architecture (Fig. 5.4) [23]. By stacking multiple electrochemical units, we can effectively multiply the output voltage to exceed 3 V—a threshold that meets the direct startup requirements of most Integrated Circuits and sensors.

This high-voltage module approach allows the system to drive BLE sensors directly, bypassing the DC-DC converter entirely. This not only simplifies the overall device footprint but also significantly improves the end-to-end energy efficiency of the harvester. These demonstrations prove that by strategically choosing between single-cell/converter pairs or high-voltage multi-leg modules, thermocells can be tailored to meet the specific power demands of a wide array of autonomous IoT applications.

5.7 Addressing Liquid Leakage: Quasi-Solid-State and Gelled Electrolytes

While purely liquid-state thermocells exhibit excellent electrochemical performance and high Seebeck coefficients, the inherent risk of electrolyte leakage poses a significant barrier to their practical, long-term deployment. This vulnerability is particularly problematic for wearable electronics, healthcare monitors, and flexible IoT sensors, where consistent mechanical stress, bending, or accidental puncture could lead to device failure and potential chemical exposure.

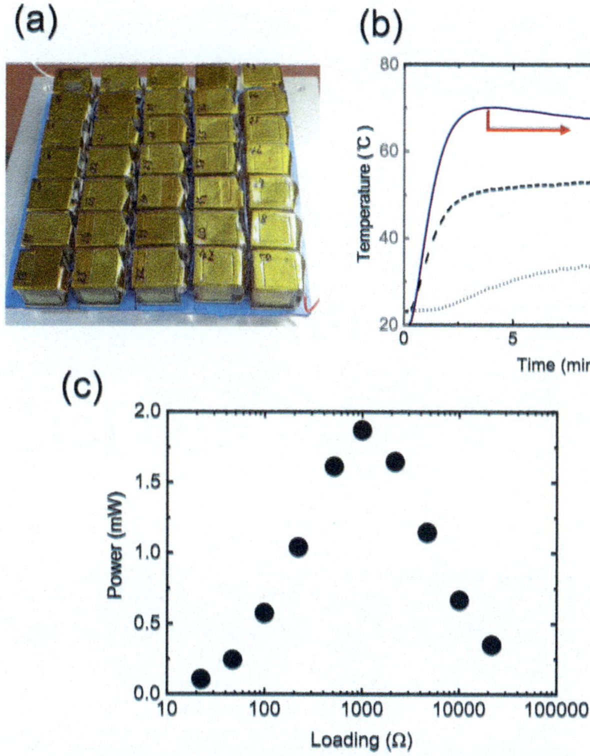

Fig. 5.4 **a** Photograph of the 35-leg thermocells connected in series using copper wire and soldering; **b** open circuit voltage plotted over time with the module placed on a 60 °C hot plate with monitored hot- and cold-side temperatures; **c** the device's power output at various loading conditions, with each loading condition measured for 1 s [23]. Copyright 2024 IOP Publishing

To overcome these structural limitations, recent research has pivoted aggressively toward the development of gelled electrolyte thermoelectrochemical cells (quasi-solid-state thermocells). By immobilizing redox-active aqueous or ionic liquid electrolytes within a three-dimensional polymer matrix—such as polyvinyl alcohol (PVA), polyacrylamide (PAAm), or cellulose-based hydrogels—researchers can create flexible, leak-proof devices. These polymer networks are carefully engineered to retain large volumes of liquid solvent, preserving the high ionic conductivity required for efficient mass transport while providing the viscoelasticity necessary for mechanical integrity (Fig. 5.5) [16, 24, 25].

However, the transition from liquid to gel is not without trade-offs. The introduction of a polymer framework can increase internal resistance and slow the diffusion of redox species, which often results in a decrease in power output during continuous operation.

Despite these challenges, the structural shift to gelled systems enables the fabrication of stretchable, conformal thermocells that can seamlessly adhere to the human

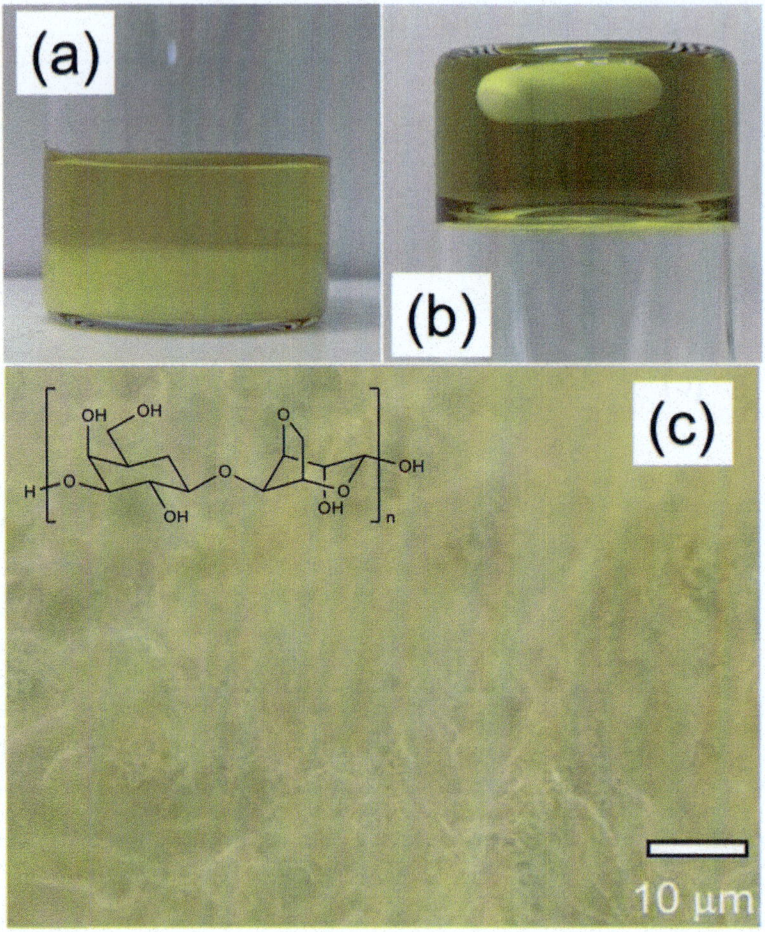

Fig. 5.5 Photographs of **a** the ferricyanide/ferrocyanide/guanidinium electrolyte solution and **b** gel prepared using 1 wt% agar with the ferricyanide/ferrocyanide/guanidinium solution. **c** Microscopic image of the gel after freeze-drying. In the inset, is reported the chemical structure of agar [16]. Copyright 2023 Wiley

body or wrap around complex industrial heat pipes. By maximizing the effective interfacial area and ensuring stable, leak-free operation, quasi-solid-state architectures bridge the gap between high-efficiency electrochemistry and the rugged requirements of real-world thermal energy harvesting.

References

1. Kuhn RJ (1930) Galvanic corrosion on cast-iron pipes. Ind Eng Chem 22:335–341
2. Reichard EC (1952) Galvanic corrosion. Chem Eng Prog 48:269–271
3. Hine RA (1957) Thermogalvanic potentials and currents at aluminum surfaces in industrial water. J Electrochem Soc 104:752–752
4. Debethune AJ, Licht TS, Swendeman N (1959) The temperature coefficients of electrode potentials. J Electrochem Soc 106:616–625
5. Katon JE, Wildi BS (1964) Semiconducting organic polymers derived from nitriles. Thermoelectric power and thermal conductivity measurements. J Chemical Physics 40:2977–2982
6. Zito R (1963) Thermogalvanic energy conversion. AIAA J 1:2133–2138
7. Burrows W (1975) Redox thermogalvanic cells for direct energy-conversion. IEEE Trans Aerosp Electron Syst 11:954–954
8. Buckingham MA, Hammoud S, Li HX, Beale CJ, Sengel JT, Aldous L (2020) A fundamental study of the thermoelectrochemistry of ferricyanide/ferrocyanide: cation, concentration, ratio, and heterogeneous and homogeneous electrocatalysis effects in thermogalvanic cells. Sustain Energy Fuels 4:3388–3399
9. Haughton-James R, Mesawang S, Buckingham MA, Taylor R, Phelan PE, Aldous L (2025) Thermogalvanic bricks: optimising large dimension thermocells for air and water valorisation. Sustain Energy Fuels 9:1165–1172
10. Hupp JT, Weaver MJ (1984) Solvent, ligand, and ionic charge effects on reaction entropies for simple transition-metal redox couples. Inorg Chem 23:3639–3644
11. Zhou H, Yamada T, Kimizuka N (2021) Supramolecular thermocells based on thermo-responsiveness of host-guest chemistry. Bull Chem Soc Jpn 94:1525–1546
12. Zhou H, Inoue H, Ujita M, Yamada T (2023) Advancement of electrochemical thermoelectric conversion with molecular technology. Angew Chem Int Ed 62:e202213449
13. Duan JJ, Feng G, Yu BY, Li J, Chen M, Yang PH et al (2018) Aqueous thermogalvanic cells with a high Seebeck coefficient for low-grade heat harvest. Nat Commun 9:5146
14. Yu BY, Duan JJ, Cong HJ, Xie WK, Liu R, Zhuang XY (2020) Thermosensitive crystallization-boosted liquid thermocells for low-grade heat harvesting. Science 370:342–346
15. Jiang L, Kirihara K, Nandal V, Seki K, Mukaida M, Horike S, Wei Q (2022) Thermoelectrochemical cells based on Ferricyanide/Ferrocyanide/Guanidinium: application and challenges. ACS Appl Mater Interfaces 14:22921–22928
16. Jiang L, Horike S, Mukaida M, Kirihara K, Seki K, Wei Q (2023) High-performance isotropic thermo-electrochemical cells using agar-gelled Ferricyanide/Ferrocyanide/Guanidinium. Global Chall 7:1900998
17. Wang W, Mukaida M, Kirihara K, Lyu L, Wei Q (2020) Poly(3,4-Ethylene Dioxythiophene)/Poly(Styrene sulfonate) electrodes in electrochemical cells for harvesting waste heat. Energy Technol 8:2070053
18. Wijeratne K, Vagin M, Brooke R, Crispin X (2017) Poly(3,4-Ethylenedioxythiophene)-tosylate (PEDOT-Tos) electrodes in thermogalvanic cells. J Mater Chem A 5:19619–19625
19. Lee JH, Jung Y, Kim JH, Yang SJ, Kang TJ (2019) Stacked double-walled carbon nanotube sheet electrodes for electrochemically harvesting thermal energy. Carbon 147:559–565
20. Zhao ZW, Nandal V, Shun A, Jiang LX, Horike S, Akaike K, Mukaida M, Kirihara K, Seki K, Wei Q (2021) Determining interfacial resistance in thermoelectrochemical cells using transmission line measurement. Appl Phys Lett 118:243907
21. Buckingham MA, Stoffel F, Zhang S, Liu Y, Marken F, Chen J, Aldous L (2022) Nanostructuring electrode surfaces and hydrogels for enhanced thermocapacitance. ACS Appl Nano Mater 5:438–445
22. Jiang L, Ebihara T, Mukaida M, Akaike K, Shimamoto K, Horike S, Wei Q (2024) Self-powered sensors utilizing single-pillar thermocells with pyrolytic graphite sheet electrodes: harvesting body heat and solar thermal energy. Mater Today Energy 45:101668
23. Jiang L, Ebihara T, Horike S, Wei Q (2024) High-voltage thermocell modules for direct device operation: eliminating the need for DC-DC converters. Mater Res Express 11:101501

all-polymer laminated modules. By addressing critical bottlenecks such as interfacial electrical resistance, these lightweight modules now achieve practical power densities in the tens to hundreds of microwatts per square centimeter. When paired with optimized power management integrated circuits, these solid-state devices successfully power Bluetooth Low Energy (BLE) sensors and facilitate the thermal recharging of secondary lithium-ion batteries, proving that an ultra-high intrinsic material efficiency is not a strict prerequisite for real-world utility.

6.2 The Liquid-State Paradigm: Thermocells and Gelled Electrolytes

Parallel to solid-state advancements, thermocells (thermogalvanic cells) offer a compelling electrochemical alternative for energy harvesting, characterized by exceptionally high Seebeck coefficients driven by temperature-dependent redox reactions. Recent innovations in material design—such as host–guest electrolytes, ultra-low-permeability flexible electrodes, and quasi-solid-state gel matrices designed to suppress thermal convection—have successfully endowed thermocells with both mechanical robustness and respectable power outputs. Furthermore, through strategic system-level engineering, the fabrication of multi-leg, series-connected modules enables these devices to directly generate the high voltages required to drive wireless communication protocols, entirely bypassing the need for intermediate DC-DC boost converters.

6.3 The Path Forward

To unlock the full commercial potential of these technologies, the next generation of research must focus on three core pillars:

- Material and Stability: For solid-state systems, the main focus should be on achieving higher charge carrier mobility alongside long-term thermal and oxidative stability of the polymers. In comparison, the focus for liquid-state thermocells requires developing ionic systems that facilitate highly efficient charge and mass transport while ensuring the oxidative stability of reduced species.
- System Integration: Deepening the synergy between device architecture and impedance-matched power electronics to maximize end-to-end energy conversion.
- Scalable Manufacturing: Transitioning from small, laboratory-scale fabrication to continuous, large-scale processes that maintain high device uniformity.

Ultimately, organic thermoelectrics and thermocells should not be viewed as direct competitors to conventional inorganic generators, but rather as specialized, complementary technologies. By capitalizing on their unique mechanical flexibility and low-cost processing, these carbon-based platforms are poised to serve as a foundational power supply for a sustainable, fully connected future.

The manufacturer's authorised representative in the EU is Springer Nature Customer Service Centre GmbH, Europaplatz 3, 69115 Heidelberg, Germany. If you have any concerns regarding our products, please contact ProductSafety@springernature.com

Printed and bound by CPI Group (UK) Ltd, Croydon, CR0 4YY
09/09/2026
02213788-0003